Engineering Materials

This series provides topical information on innovative, structural and functional materials and composites with applications in optical, electrical, mechanical, civil, aeronautical, medical, bio- and nano-engineering. The individual volumes are complete, comprehensive monographs covering the structure, properties, manufacturing process and applications of these materials. This multidisciplinary series is devoted to professionals, students and all those interested in the latest developments in the Materials Science field.

More information about this series at http://www.springer.com/series/4288

Inamuddin · Rajender Boddula ·
Abdullah M. Asiri
Editors

Black Phosphorus

Synthesis, Properties and Applications

Springer

Editors
Inamuddin
Department of Chemistry
Faculty of Science
King Abdulaziz University
Jeddah, Saudi Arabia

Department of Chemistry
Faculty of Engineering and Technology
Aligarh Muslim University
Aligarh, India

Abdullah M. Asiri
Department of Chemistry
Faculty of Science
King Abdulaziz University
Jeddah, Saudi Arabia

Rajender Boddula
CAS Key Laboratory of Nanosystem
and Hierarchical Fabrication
National Center for Nanoscience
and Technology
Beijing, China

ISSN 1612-1317 ISSN 1868-1212 (electronic)
Engineering Materials
ISBN 978-3-030-29554-7 ISBN 978-3-030-29555-4 (eBook)
https://doi.org/10.1007/978-3-030-29555-4

This Springer imprint is published by the registered company Springer Nature Switzerland AG
The registered company address is: Gewerbestrasse 11, 6330 Cham, Switzerland

Preface

Two-dimensional materials, for example, graphene and transition metal dichalcogenides (TMDCs), have recognized huge consideration in the earlier decade. Therefore, there are numerous fundamental science motivations to create layered two-dimensional (2D) materials. Phosphorus is standout amongst the most abundant components found on the earth. Black phosphorus (BP) is the most stable allotrope of phosphorus, having multiple layers with two-dimensional structures. Layered BP and its monolayer flatland material (phosphorene) pulled enormous research enthusiasm since the discovery of BP-based field-effect transistors early in 2014. Since then, black phosphorus is the cutting-edge semiconductor material considering its bandgap, anisotropy, high carrier mobility, phenomenal physical and electrical properties. Black phosphorus possesses unique properties that can bridge the gap between graphene and TMDCs for various potential applications. Therefore, awareness and knowledge about black phosphorous materials with conceptual understanding are essential for the advanced materials community.

Black Phosphorus: Synthesis, Properties and Applications aim to explore down to earth applications in the fields of biomedical, environmental, energy and electronics. This book provides an overview of the structural and fundamental properties, synthesis strategies and various applications of black phosphorus. This book will help the readers to solve fundamental and applied problems faced in the field of black phosphorous applications. The book content incorporates industrial applications and will fill the gap between the laboratory scale to practical applications. This book will target industrialists, scientists, university professors, lecturers, researchers, Ph.D. and master students working in the field of material science, semiconductor technology, energy and environmental science.

Jeddah, Saudi Arabia/Aligarh, India — Inamuddin
Beijing, China — Rajender Boddula
Jeddah, Saudi Arabia — Abdullah M. Asiri

The original version of the book was revised: The book editors' affiliation text has been updated. The correction to the book is available at https://doi.org/10.1007/978-3-030-29555-4_10

Contents

Functionalization and Doping of Black Phosphorus

Mehdi Ghambarian, Zahra Azizi and Mohammad Ghashghaee

Abstract Black phosphorus (BP), a new rising star in the post-graphene era, has been extensively studied from early 2014 onward. To take full advantage of its potential, much research is rapidly generated on the modification of BP-based nanostructures via functionalization, decoration, and doping. The fast-growing research in this area has led to significant trends in the fast-evolving field of 2D nanomaterials over a wide range of applications including field effect transistors, diodes, phonon detectors, biomedicine, digital circuits, sodium- and lithium-ion batteries, sensors, photocatalysis, electrocatalysis, thermoelectric materials, memory devices, and so forth. This chapter is aimed to present a state-of-the-art overview of the advancements of the field through the modifications mentioned above from both theoretical and experimental points of view.

Abbreviations

2D	Two-dimensional
3D	Three-dimensional
AFM	Atomic force microscopy
AIBN	Azodiisobutyronitrile
ALD	Atomic layer deposition
BCS	Bardeen–Cooper–Schrieffer
BP	Black phosphorus

M. Ghambarian
Gas Conversion Department, Faculty of Petrochemicals, Iran Polymer and Petrochemical Institute, P.O. Box 14975-112, Tehran, Iran

Z. Azizi
Department of Chemistry, Karaj Branch, Islamic Azad University, P.O. Box 31485-313, Karaj, Iran

M. Ghashghaee (✉)
Faculty of Petrochemicals, Iran Polymer and Petrochemical Institute, P.O. Box 14975-112, Tehran, Iran
e-mail: m.ghashghaee@ippi.ac.ir; ghashghaee.m@gmail.com

Inamuddin et al. (eds.), *Black Phosphorus*, Engineering Materials,
https://doi.org/10.1007/978-3-030-29555-4_1

BPQD	BP quantum dot
DCD	Differential charge density
DFT	Density functional theory
DMS	Dilute magnetic semiconductor
DV	Divacancy
ESE	Emulsion solvent evaporation
F4-TCNQ	Tetrafluoro tetracyanoquinodimethane
FET	Field effect transistor
GGA	Generalized gradient approximation
IR	Infrared
LDOS	Local density of states
MD	Molecular dynamics
MGPT	Mineralizer-assisted gas-phase transformation
MV	Monovacancy
NDR	Negative differential resistance
NIR	Near-infrared
NVM	Nonvolatile memory
PA	Photoacoustic
PBC	Periodic boundary conditions
PDDA	Poly dimethyldiallyl ammonium chloride
PEG	Polyethylene glycol
PGE	Photogalvanic effect
PLGA	Poly(lactic-co-glycolic acid)
PVR	Peak-to-valley ratio
RP	Red phosphorus
SPR	Surface plasmon resonance
TCDD	Tetrachlorodibenzo-*p*-dioxin
TCNQ	Tetracyano-*p*-quinodimethane
TEM	Transmission electron microscopy
TM	Transition metal
TMD	Transition metal dichalcogenide
TTF	Tetrathiafulvalene
VBM	Valence band maximum
VD	Vacuum deposition
vdW	van der Waals
WI	Wet impregnation
WP	White phosphorus

1 Introduction

Two-dimensional (2D) nanomaterials are at the center of a huge amount of scientific contributions owing to their exceptional capabilities for intriguing new prospects. Black phosphorus (BP) nanostructures composed of only one element have earned their place among 2D semiconductor nanomaterials owing to their high optical/UV absorption, high carrier mobility, and other capabilities of particular interest to different disciplines [1–4]. As of today, a total of five polymorphs and several amorphous forms of phosphorus have appeared experimentally [5], while several other phases have also been investigated theoretically [6]. Although BP is thermodynamically the most stable high-density phase allotrope of phosphorous materials [5], it appeared experimentally exfoliated several centuries after the historical discovery of the element in ancient China, where the combustion of P_2H_4 was described [7].

Nevertheless, the earliest discovered molecular form of this element, called white phosphorus (WP), was isolated in 1669 by Hennig Brand. Later in 1914, Percy Bridgman introduced the phase transition of WP to BP under high pressure [7]. In contrast to graphene, phosphorene and the few-layer BP materials present anisotropic orthorhombic structures that are ductile in one in-plane direction while being stiff along another. This diversity brings about unusual mechanical, optical electronic, and transport properties that expound the anisotropy of the matrix [1]. The role of functionalization, decoration, and doping in providing tailor-made properties of the BP-based devices are surveyed enormously [1, 8].

Nonetheless, BP attracted little attention as a semiconductor due to its narrow bandgap (~0.3 eV) and the challenges in controlling its quality [9] until 2014, when some publications revealed the unusual properties (a direct bandgap and high carrier mobility) of monolayer and thin-layer BP [10]. These findings triggered the beginning of a new trend in the research on BP and especially what came to be named phosphorene (monolayer BP) [10] in analogy to graphene [1], which sometimes is also designated loosely to few-layer BP materials. It is worth mentioning that the IUPAC designation would be 2D phosphane as the material has no sp^2 bond (Fig. 1) [1, 11].

Although the topic at hand is relatively new, several reviews have already appeared in the field [12], covering material fabrication, properties, and applications. Nevertheless, we are unaware of any systematic review of the fast-growing research on the three modification routes for different applications. The peculiar structure of BP sets it aside from other widely explored 2D semiconductors such as graphene and transition metal dichalcogenides (TMDs). In the past few years, modern techniques adapted to the processing of 2D materials have enabled the isolation and identification of monolayer and few-layer BP. Some of the remarkable features of phosphorene include high electron and hole mobilities, high thermoelectric ZT, a negative Poisson's ratio, and superconductivity [1].

Currently, the main bottlenecks in the research of phosphorene include limited availability and a high tendency for degradation. The former can only be tackled via more substantial investment for large-scale production of the material, which in turn will be fueled as the potential advantages are uncovered progressively. The second obstacle. i.e., the fast degradation when exposed to ambient conditions can be circumvented such that it will not be a limiting factor in two ways. The first viable and reliable technique is encapsulation as inspired by the standard practice in the organic light-emitting diode industry. All known encapsulation methods, such as the atomic layer deposition (ALD) of alumina, PMMA encapsulation, and encapsulation

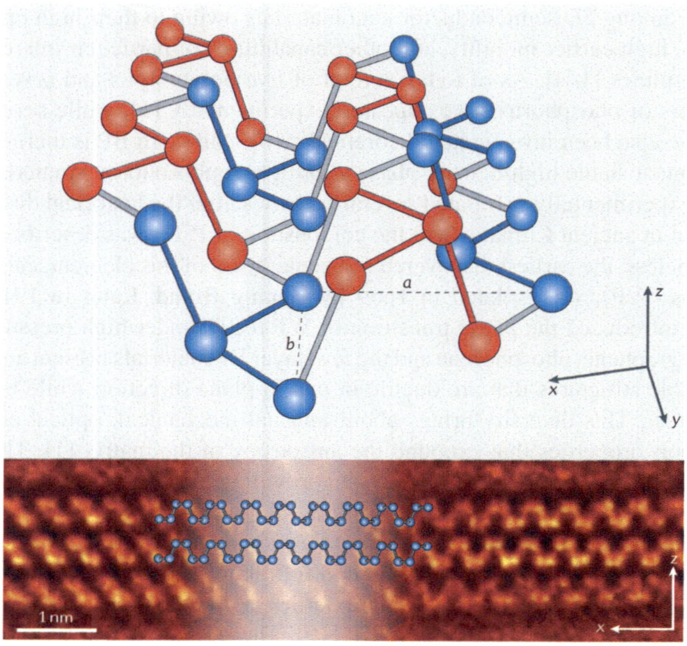

Fig. 1 Base-center orthorhombic matrix of bulk BP, with the lattice constants *a* and *b* being indicated in the top panel. An aberration-corrected transmission electron microscopy (TEM) image is depicted in the lower panel with an overlaid ball-and-stick model [1, 3]

in 2D stacked heterostructures have been applied to monolayer and few-layer BP with auspicious outcomes. The second aspect is that the same degradation ability can be exploited in some areas of interest, such as trace O_2 sensing [1].

Despite the limitations mentioned above, the few-layer BP can preserve high crystal quality after isolation as implied from quantum oscillations. A variety of phosphorene-based devices, to name a few, the FETs and gas sensors have appeared so far with performances that surpassed the initial expectations. For instance, the BP-based FETs have exhibited high mobility, a promising current saturation, and a balanced on/off ratio, particularly for the radio-frequency purposes. Furthermore, the tunable bandgaps from 0.3 eV (bulk BP) to 2.0 eV (phosphorene monolayer) [12] covers the broad spectrum from visible light to infrared (IR) radiation, while also encouraging for the photodetection. Thanks to the anisotropic properties of BP monolayer, the angle-resolved transport applications are also of interest. The same capability combined with IR and Raman spectroscopy may be utilized to determine the direction of the BP crystals, thus enhancing the quality of device fabrication [1, 2].

BP offers exciting opportunities for the band structure engineering and, therefore, the design of 2D heterostructures. However, many theoretically estimated properties of the BP monolayer are still in need of further verification, which include the optical and transport bandgaps, the governing factors that control the electron and hole mobility, the identification of low-angle edge vacancies, the direct band structure of the few-layer BP, the identification of higher exciton modes, the detection of the upper limit for the phonon-restricted carrier mobility, and the observation of magnetism and superconductivity [1].

This chapter reviews the significant advances in the improvement and modification of BP using functionalization, doping and decoration. Several authors have employed the terms *functionalization* and *doping* interchangeably. Modifications through surface *adsorption* of different species have also been called functionalization in some publications. More complicating in this regard is that no clear distinction is often made between the *substitutional* doping as the common perception of such in-plane changes more widely applied in materials science and the so-called *adatom* doping, which is frequently used in this area. Here, we prefer to adopt the more distinctive term *decoration* for the latter, which addresses all non-substitutional interactions through adsorption of an atom or molecule. This type of modification is called physical functionalization in some tests. Eventually, we note that a sharp demarcation between these definitions would be difficult in many cases in which the classification should be regarded as almost tentative.

2 Methods and Applications

Even though no exact classification is available for the modification methods, one may name a few. Many relevant works make use of theoretical methods, e.g., density functional theory (DFT), ab initio, and molecular dynamics (MD) simulation. Most of the models have been defined under periodic boundary conditions (PBC) with varying unit cell sizes (Fig. 2).

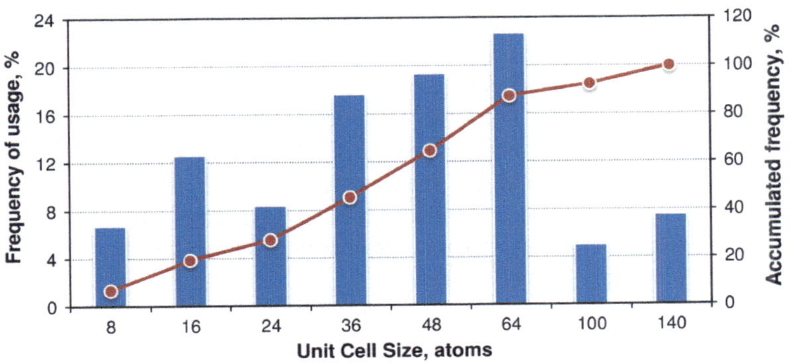

Fig. 2 Variety of the unit cell sizes in the PBC calculations of the BP systems (evaluated with about 120 publications)

A variety of experimental techniques have been employed to the BP modification. These methods include atomic layer deposition (ALD) [13–16], controlled wet impregnation (WI) [17–23], sputtering [24], thermal evaporation [22, 24], vacuum deposition (VD) [25, 26], electrochemical exfoliation with synchronous functionalization [27], mineralizer-assisted gas-phase transformation (MGPT) [28], emulsion solvent evaporation (ESE) [29], in situ polymerization [30], ball-milling [31–34], and spin coating [35–37].

BP materials are modified through doping, decoration, and functionalization for numerous foreseeable applications, that are briefly reviewed in the following.

(1) Electronic devices, which include field effect transistors (FETs), spintronic devices, thermoelectric materials, and optoelectronic devices. FET is a three terminal device that applies an electric field to control the current flow through a device [38–42]. Similarly to the electronic devices that apply the electrical charge of an electron to encode data, the spintronic or spin electronic devices employ another fundamental property known as the electron spin, which is the intrinsic angular momentum of the electron, to transmit, process and store information [40–42]. A thermoelectric device is able to convert heat flow into electrical energy. Owing to their anisotropic thermal and electronic transport properties, BP nanostructures may form an ideal choice for thermoelectric applications [40–42]. Optoelectronic devices (detectors, modulators, switches, and amplifiers), also called photonic devices are essential tools in many modern high-performance applications, such as biomolecule sensing, optical communications, and data storage and retrieval. These devices also include photonic elements usually found in fiber communication and data storage systems [41].

(2) Energy storage, e.g., as sodium and lithium-ion batteries and solar cells. Lithium-ion batteries are a family of batteries in which the Li cations shift from the negative electrode to the positive one during the discharge and move back during the recharging [3, 40–44]. Solar energy is converted into electrical energy using a photovoltaic solar cell, a semiconductor device in which the sunlight energy within a certain range of wavelength can be absorbed to produce free electrons on one side and holes on another [45].

(3) Catalysis, which includes heterogeneous and homogeneous catalysis, biocatalysis, and photocatalysis. The latter term describes a process in which the rate and/or the outcome of a photochemical reaction is influenced by the surface of a solid catalyst, normally a semiconductor, which is defined by a filled valence band and an empty conduction band. Examples include water splitting, N_2 reduction to NH_3, sensitizing of a light-induced redox reaction, and more [41, 46, 47].

(4) Sensors and Sorbents. Modifications can improve the parent material for the adsorption/capture of different species. Sensors are devices for converting one type of energy into electrical energy, which along the acquisition and processing instrumentation enable the displayed detection of different analytes. Sensors also include optoelectronic devices/photonic sensors for the monitoring and detection of humidity, metals, and different gases. In a direct sensor, e.g.,

the physiological electrodes, the measured variable is electrical, or it can be directly converted into electricity. Sensors can also be classified into active and passive devices. A passive one, e.g. photodiode, requires no external power for the acquisition of the electric signal. The transduction of chemical sensors falls into three classes depending on their modes of measurement, which include changes in (i) electrical/electrochemical properties, (ii) physical properties, and (iii) optical absorption. Optical sensors have the capability to quantify different characteristics of light including intensity, wavelength, frequency, and polarizability [40–42, 44, 48]. The stable and tunable conductance of BP relative to the other 2D materials have enabled the construction of highly sensitive few-layer BP sensors [49].

(5) Composite materials. The remarkable mechanical properties of BP nanostructures can be an advantage when integrated with other materials; this enables the design of ultra-high strength composites over a wide range of fields [41]. Composite materials consist of two or more chemically/physically different phases that have been separated by a distinct interface, but wisely integrated for a new functional/structural behavior. Commonly, the composite materials encompass three phases including (i) the matrix or the continuous phase, (ii) the filler/reinforcement phase, and (iii) the interface [50].

(6) Biomedical applications, viz. for biosensing, photoacoustic imaging, drug delivery, and (photothermal and photodynamic) cancer therapies [49]. A biosensor detects biomolecules such as enzymes, nucleic acids, ap-tamers, whole cells, and so forth [51]. FET-based immunosensors detect an alteration appeared in the channel conductivity, which is driven by the electric field of the medium. The source–drain channel conductivity is proportional to its charge carrier density [52]. Electrochemical biosensors convert biochemical information, e.g. the analyte concentration into a detectable signal, such as current or voltage [53]. Optical biosensors measure the light absorbed or emitted as a consequence of a biochemical reaction based on different optical techniques such as absorption, luminescence, fluorescence, and surface plasmon resonance (SPR) [51]. Fluorescent biosensors enable imaging and quantifying the enzyme/protein activity and cellular signals [54]. Another biomedical application field is the promising noninvasive cancer imaging. Photoacoustic (PA) imaging offers high image contrast in the region of 5–6 cm deep in the tissues for background-free detection of the tumor. Graphene oxide alternatives have been sought to an improved photodetection while maintaining high stability. BP has been employed in cancer phototherapy, with high therapeutic efficiency and minimal invasiveness [49].

Figure 3 presents how often the three modification routes have been employed to the application mentioned above. As can be seen, all three modifications have mainly shown advantages directed to the applications in the electronic devices. However, the biomedical applications, catalysts, and sensors have been the next targets of BP modification through functionalization, decoration, and doping, respectively. In total,

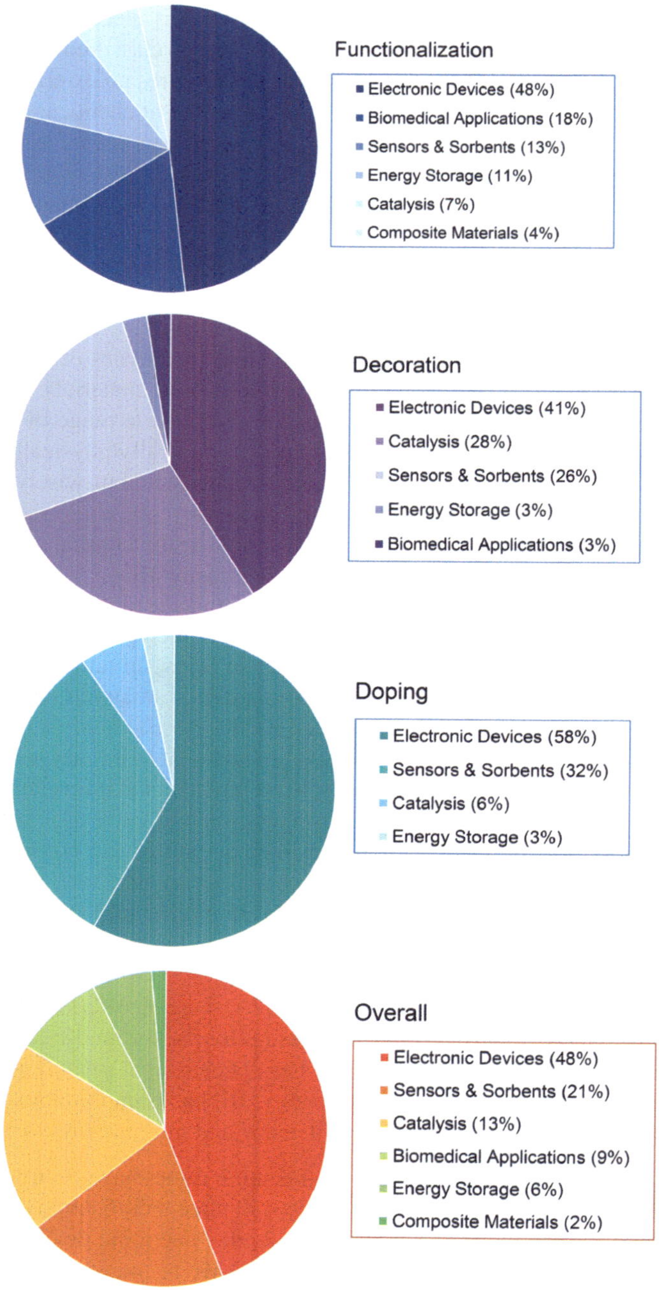

Fig. 3 Statistical distributions for the frequency of usage of different modification methods in different application areas leading to 2019 (according to Scopus)

the modification routes were applied to different fields from the lowest to the highest frequency of usage as follows: composite materials, energy storage, biomedical applications, catalysis, sensors, and electronic devices.

3 Functionalization of Black Phosphorus

This section summarizes the attempts into the improvement of BP nanomaterials using different functional groups incorporated for different purposes. It is quite reasonable to assume that the chemical (covalent) bindings occur preferentially at the pre-existing basal plane defects or grain boundaries [55]. Such chemical functionalization endows BP with higher stability against oxidative degradation along with the changes in the electronic properties [55]. In contrast to the wonderful physical properties of BP which have been intensively discussed [56], the chemistry of BP has remained less investigated [55]; only recently, several non-covalent [57] and covalent [58] functionalization scenarios have been reported [55]. For instance, functionalization, through the addition of electrophiles such as carbenes, O, nitrenes, and Lewis acids has recently been reported [55]. DFT calculations [59] have shown that the dissociative hydrogenation of phosphorene using the gas-phase H_2 molecules (with an energy barrier of 2.54 eV) provides a remarkable quasiparticle band gap of 2.29 eV, while the fully hydrogenated material was dynamically unstable. The models suggested the decomposition of the BP monolayer into weakly bonded 1D chains in the zigzag direction.

Phosphorene oxidation with a 50% coverage in another theoretical study revealed different exothermic possible configurations, three of which involved interstitial oxygen bridges. Electronic structure calculations signified a corrected band gap modulation in the range of 1.2–2.9 eV using the GW method. The analysis of both Young modulus and Poisson ratio revealed that the mechanical response of all conformers was highly direction-dependent, thus indicating that the new derivatives were incompressible materials [60].

The appraisal of the phonon dispersion curves predicted that 2D phosphorene oxide could be stable at ambient conditions in both stoichiometric and non-stoichiometric configurations [61, 62]. The nature of the band gap, however, depends on the degree of functionalization, such that indirect and direct gaps were predicted for the non-stoichiometric and stoichiometric phosphorene oxide materials [61].

The BP vacancies and grain boundaries have a stronger O_2 affinity (with about 5000 times higher oxygen dissociation rates) than the perfect lattice sites. Charge transfer assessments have indicated that the O_2 molecule is a potent electron scavenger that increases the hole carriers and transfers the lone-pair electrons of BP to the $2\pi^*$ antibonding orbitals of the oxygen molecule [63, 64]. Disordered amorphous-like BP structures are formed at high concentrations of oxygen [59]. Similar studies with the many-body perturbation theory have explained [65] the influence of oxygen arrangements and excitons on the optical properties of phosphorene. Unlike pristine phosphorene, the absorption spectrum of the oxidized structures was extended along the solar (light) spectrum, with higher absorption coefficients in the dangling bonds. The authors found that the first exciton state in all PO conformers was optically dark, making them appropriate for long-lived applications. Full fluorination of phosphorene through the low-energy (0.19 eV) dissociative chemisorption of F_2 leads to a dynamically stable semiconductor with an indirect band gap of 2.27 eV. However, the monolayer structure was decomposed into narrow PF chains [59, 66].

Ball-milling of bulk BP powders in the presence of a LiOH additive without using any noble metal cocatalyst led to the OH functionalization of BP. Metal-free photocatalysts for hydrogen evolution are preferred over the conventional metal-based inorganic photocatalysts as nonmetal elements are normally less costly, more earth-abundant, and eco-friendly. The light photocatalytic H_2 evolution rate of few-layer BP nanostructures approached 512 $\mu molh^{-1}g^{-1}$, which is comparable to that of graphitic carbon nitride (g-C_3N_4) and over one order of magnitude higher than that of bulk BP [31].

In the presence of high-concentration N_2H_2, single-sided or double-sided (=NH) functionalization of BP monolayer occurs at ambient conditions to form imine-functionalized phosphorene structure [62]. Urea-assisted ball milling of BP has also created NH_2-functionalized nanosheets with a hydrogen evolution turnover frequency of 3.21 s^{-1} at an overpotential of 290 mV, markedly outperforming the bulk BP. The electrocatalytic activity of the functionalized material in alkaline medium was notably higher than that in an acidic electrolyte, thus rendering as a promising metal-free hydrogen evolution catalyst [32]. The intrinsically acidic BH_3 functional favors the periodic mono-hapto anchoring at the BP atoms. The resulting adducts were compared to similar molecular P donors. Even though the P lone pairs of the BP surface have a limited donor capability and the adducts have never been experimentally demonstrated, the new hapticities appeared to be reasonably stable according to the structural, spectroscopic, electronic, and energetic data [67].

Unencapsulated BP FETs undergo chemical degradation at ambient conditions, which largely increases the threshold voltage after 6 h, followed by a dramatic decrease in the current on/off ratio after 48 h. Passivation of BP through the atomic layer deposition (ALD) of AlO_x overlayers could successfully control the BP degradation, thus enabling high on/off ratios (about 1000) and hole mobilities (about 100 $cm^2V^{-1}s^{-1}$) for an extended period (over two weeks) under ambient conditions (Fig. 4) [14].

A covalent functionalization with fragments that incorporate CO ligands, i.e., $(CO)_5Mo$, $(CO)_3Mo$, and $(CO)_4Ru$, $(CO)_2Ni$, and $Cl_2(CO)Pt$ were examined theoretically. The concept of isolobal analogy was found to be critical for the modeling of the possible sets of external coligands in the functionals of a varying number of vacant σ hybrids. All of the interactions were found to be of the acid–base Lewis type, stemming possibly from the compactness of the lone pair electrons of BP that contribute to the valence band. The same discussion held for $(CH_3)_2Ni$ in which the quasi-square planar coordination of the d^8-Ni(II) on the BP monolayer gave rise to an increased band gap relative to the bare BP [67] (Figs. 5 and 6).

Surface coordination of BP with a titanium sulfonate ligand (TiL_4) fragment led to an excellent air and water stability for an extended period of 72 h, thereby

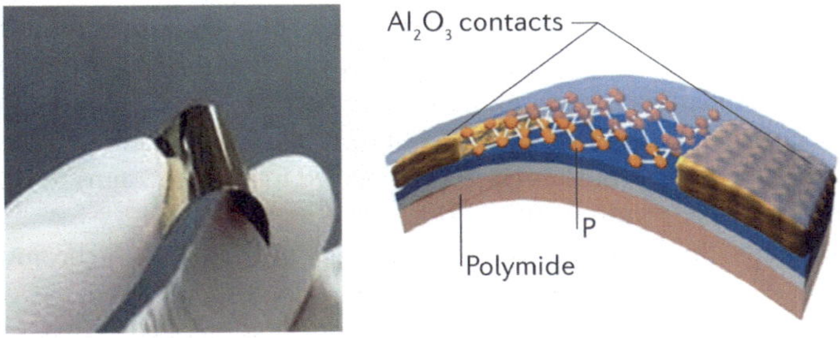

Fig. 4 Photograph and schematic diagram of a flexible phosphorene-based transistor prepared through encapsulation by AlO_x using ALD [1]

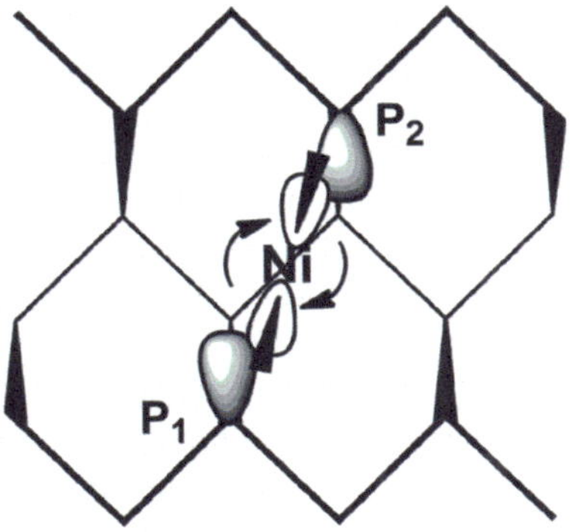

Fig. 5 Proposed rearrangement of the $L_2Ni(II)$ fragment across a phosphorous channel for higher σ overlap [67]

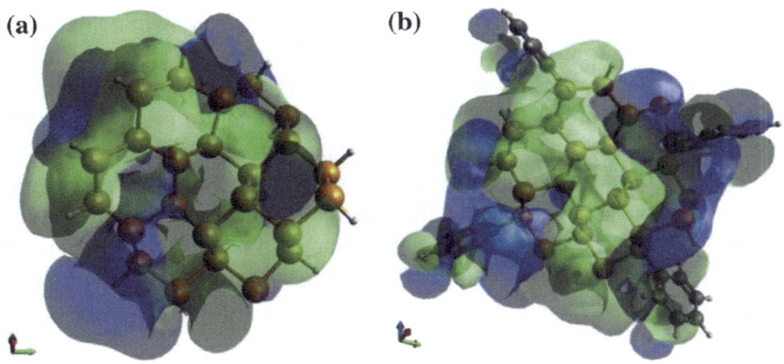

Fig. 6 LUMO orbitals for **a** $P_{24}H_{12}$ and **b** 4Benz$P_{24}H_8$ [68]

substantially extending its lifetime and spurring its effective application in killing the cancer cells [19]. In a theoretical study, functionalization of a BP quantum dot (BPQD) that contained 24 P atoms and 12 H atoms with different aromatic fragments, such as benzene and anthracene showed interesting optical features. For instance, BPQD functionalization with four benzene rings led to a large (7 nm) redshift of the first transition in the absorption spectra [68].

In another study, the exceptionally high reactivity of BP toward *S*-(bromomethyl)-ethanethiolate (C_3H_5BrOS) led to the direct formation of P–C bonds, as confirmed by the XPS and Raman spectroscopy [69]. Functionalization of BP with Cs_2CO_3 was found to strongly donate electrons to BP such that the electron mobility of BP was greatly improved to ~27 $cm^2V^{-1}s^{-1}$ thus enhancing the electron transport behavior and responsivity of BP photodetectors to 2.56 without degrading. Similar findings were obtained with MoO_3 decoration that bestowed a tremendous hole-doping power as evinced by the high interface charge transfer identified by in situ photoelectron spectroscopy [26].

Covalent functionalization of BP with methoxybenzene, nitrobenzene, and phenylenevinylene revealed that BP modification leads to a smooth energy band within the original bandgap, thus partially decreasing the hole mobility of pristine BP [70]. Covalent functionalization of BP with 4-azidobenzoic acid led to five-coordinated bonding of P atoms, which completely passivated the reactive BP, thus improving its ambient stability [21]. Prior to this finding, the passivation routes were limited to aryl diazonium and nucleophilic additions that furnished P–C and P–O–C single bonds [21, 58]. In such instances, the remaining unpaired electrons in the P atom was an obstacle to effective passivation. However, the azide passivation could lead to the unprecedented formation of P=N bonds, garnering new insights into the nitrene chemistry and promises for potential applications of BP [21].

The ability of diaryliodonium salts to arylate both P and O nucleophiles is preferred over diazonium salts which only react at the P sites. Meanwhile, the reasonably low reactivity of the aryliodonium salts makes them compatible with the room-temperature passivation of BP with no escalated oxidation. Furthermore, the aryl modification using iodonium salts provides a new pathway for the covalent BP modification and subsequent binding of electron donating and withdrawing substituents to the aryl group while the ridged oxide formation is restricted [22]. Significant electron transfer from the BP flakes to the organic fragment occurred upon the non-covalent functionalization of BP with 7,7,8,8 tetracyano-p-quinodimethane (TCNQ), in which the positive charge on the upper BP layer was stabilized by the lower layers. Passivation with the perylene diimide drastically enhanced the BP resistance against oxygen degradation [71].

Improved stability in either water or physiological media was also observed with functionalization of BPQD with polyethylene glycol (PEG) for effective photothermal cancer therapy [23, 72, 73]. Analogously, hydrophobic poly(lactic-co-glycolic acid) (PLGA) was found to not only isolate BPQD from water and oxygen to increase its photothermal stability (e.g., for 24 h in a phosphate-buffered saline medium), but also control its degradation rate, thus producing biodegradable

BPQDs/PLGA nanospheres for nontoxic tumor targeting capability [29]. Modification of BP nanoflakes with the carbon free radicals of azodiisobutyronitrile (AIBN) molecules can offer high stability in air and aqueous solution while maintaining reasonable optical properties [74]. Similar findings of promising passivation effects have been reported with octadecyltrichlorosilane [75], poly dimethyldiallyl ammonium chloride (PDDA) [76], benzyl viologen [77], tetracyano-*p*-quinodimethane (TCNQ) [78, 79], tetracyanoethylene [79], tetrathiafulvalene (TTF) [78, 79], and 2,3,5,6-tetrafluoro-7,7,8,8-tetracyanoquinodimethane (F4-TCNQ) [36]. A complex functionalized system called Au/PDDF-*g*-BP/ITO that involved a highly soluble conjugated polymer covalently functionalized BP material offered nonvolatile rewritable memory performance, with an on/off current ratio of 10^4 and on/off voltages of + 1.95/–2.34 V [80].

4 Black Phosphorus Decoration

Decoration of BP with different adatoms can tune its properties for a variety of applications. This section provides an organized summary of the advancements in this area. Overall, three different adsorption sites are plausible. Some authors recognize two types for the bridge stabilization [81].

Theoretical estimates of binding energy and cohesive energy have shown that Li, Na, K, Rb, and Cs alkali metals from group IA can effectively decorate black phosphorene without any clustering. Most of the decorated structures demonstrated improved capability for the reversible CO storage or CO removal [82]. BP monolayer strongly bonds with all mentioned adatoms while keeping its structural integrity with the most considerable deformations of only 0.1–0.2 Å [81, 83]. Such doped nanomaterials present n-type behavior [84–86] with low ionization energy impurity states induced into the band gap [85], and are potential sensors for molecular hydrogen [87]. Li and Na doping induced a metallic state in black phosphorene [88]. As the size of the alkali metal increases, the absorption coefficient is decreased while shifting toward the visible region [89]. Inexpensive Li decoration leads to the strong adsorption of CO_2 with an adsorption energy of 0.376 eV [90].

Intercalating the Li atoms in the BP bilayer can transform the original direct-gap semiconductor to an intrinsic Bardeen–Cooper–Schrieffer (BCS) superconductor, in which the electron occupation of the Li-derived band is diminutive. Increasing the number of Li atoms increases both electron–phonon coupling and metallicity [91]. Potassium decoration gives rise to an apparent bandgap reduction and improved electron mobility by over one order of magnitude [92].

Phosphorene decorated with Ca, Sr, and Ba from group IIA could be applied to the reversible CO capture [82] and hydrogen storage [87, 93, 94]. Calcium-decorated BP monolayer shows n-type behavior [94]. Calculated energy barriers have demonstrated that the diffusion of the Be, Mg, and Ca adatoms on a BP monolayer is quite anisotropic owing to its puckered structure [85]. Other investigators mention that Ca and Mg adatoms give rise to mid-gap states and, therefore, are not going to be effective electron donors [84]. BP decorated with Mg^{2+} is regarded as a high-performance transistor [18]. The assessment of adsorption and cohesive energies suggested that the Mg adatom prefers three-dimensional (3D) island growth and coarsening [88]. Phosphorene decoration with the Ca and Sr adatoms were relatively promising for CO_2 capture [95].

Among the elements of group IIIB, the adsorption of La on black phosphorene could successfully modify it for adsorption of multiple CO molecules [82]. Sc prefers an H site for adsorption on phosphorene [85]. Both Sc and La demonstrated adsorption energies that fell within the ideal energy window of 0.2–0.6 eV for practical hydrogen storage applications [87]. Black phosphorene decoration with Ti and Zr from group IVB led to a strong coupling between the metal-decorated phosphorene and CO molecules [82]. Similarly, Ti-decorated phosphorene turned out as a promising candidate for CO_2 sensing and removal [95]. Titanium has been experimentally used as an electrode in few-layer phosphorene transistors [96, 97]. The adsorption energies of metal adatoms such as Ti on black phosphorene were more than twice greater than on graphene [81]. Such nanostructures are potential materials for hydrogen capture [87, 93]. The adsorption of Ti adatom provides a spin-polarization with a magnetic moment of 1.87 μ_B in a $3 \times 3 \times 1$ supercell [84]. Decoration with Hf (on H sites of BP) was beneficial for enhanced thermal stability of the resulting transistor [13].

For the elements of group VB, the cohesive and binding energies of the adatoms are quite large (Nb > V) with little distortion [81, 82]. Among many others, the vanadium-decorated phosphorene has proven the highest capability for an efficient binding with CO_2 in which the preferred configuration entailed an oxygen atom of CO_2 pointing to the phosphorene surface [95]. Relatively strong adsorption toward hydrogen was also reported for the V-decorated BP structures [87]. Adsorption of Nb induces the magnetic moment of about 1.00 μ_B [85].

Among several TM adatoms, Cr and Mo from group VIB demonstrated the highest ratio of the adsorption energy to the bulk cohesive energy, which indicated that the Cr and Mo islands have higher thermal stability against island coarsening. Therefore, these two systems, particularly the Mo-decorated one, can be used for potential surface-supported catalysis applications [88]. Cr adatom has also shown high potential for CO_2 capture [95]. BP decoration with Mn and Tc from group VIIB can also enhance its adsorption ability toward H_2, CO, and CO_2 with little distortions [81, 82, 87, 95].

Decoration of BP monolayer with Fe, Ru, Os, Co, Rh, Ir, Ni, Pd, and Pt from group VIIIB leads to a range of large binding energies [82, 85, 96–99]. Fe and Co induce magnetic moments of 2 and 1 μ_B, respectively, thus making the BP monolayer a magnetic semiconductor [98]. Some of the adatoms mentioned (Ni, Co, and particularly Fe) significantly enhanced the interaction between CO_2 and the BP monolayer. Some others (Pd and Pt) were almost ineffective in this regard [95]. Instead, Pt greatly facilitates hydrogen storage and dissociation [93]. Spin-polarized band structures are obtained with Fe and Co decoration (Fig. 7) [84, 86, 88]. Except for Pd, all other TM adatoms mentioned above could elongate the O–O

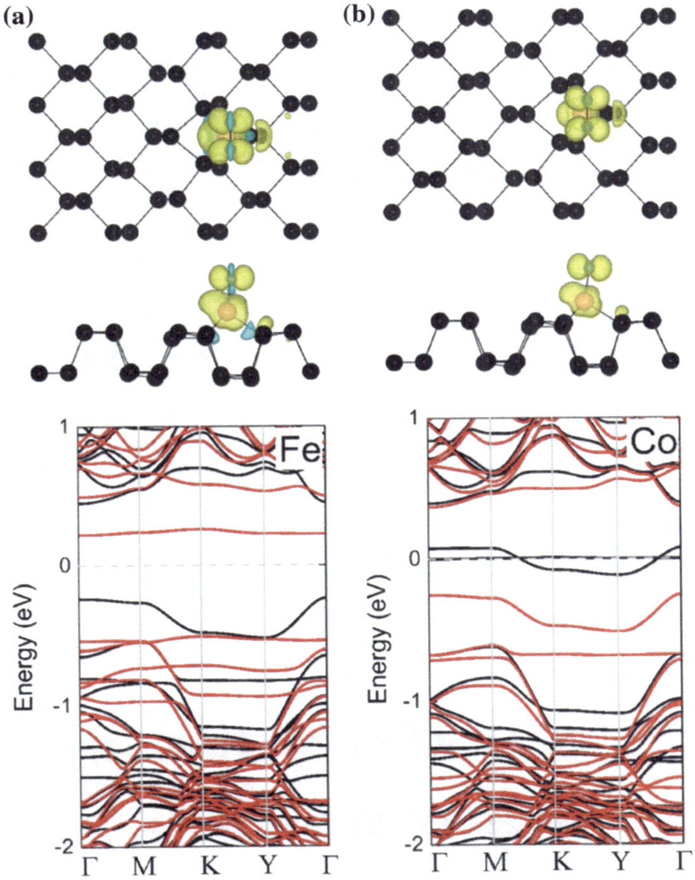

Fig. 7 Superimposed spin-polarized charge density plots of **a** Fe-decorated and **b** Co-decorated phosphorene systems (illustrated in the top row) with isosurface values of ±0.002 e/Å^3, and the corresponding band structures (depicted in the bottom row); the black and red balls indicate P and TM atoms, respectively; the yellow and cyan areas denote the up and down spins, respectively; the black and red lines in the band structures correspond to the up and down spin channels, respectively; the grey zero dashed line represents the Fermi level [98]

bond as a critical measure of their applicability to the catalytic oxidation of the CO molecule (Fig. 8) [98]. The Fe-decorated structure can be a potential candidate as a dilute magnetic phosphorene semiconductor as no clustering takes place during the modification [88]. BP decoration with Fe^{3+} is also beneficial in the development of modern FETs [18].

Theoretical data have indicated that the Pt-decorated structure presents the strongest (−1.101 eV) adsorption capacity for H_2S removal (Fig. 9). However, NH_3 and HCN can be eliminated more effectively by the Ni-decorated BP monolayer,

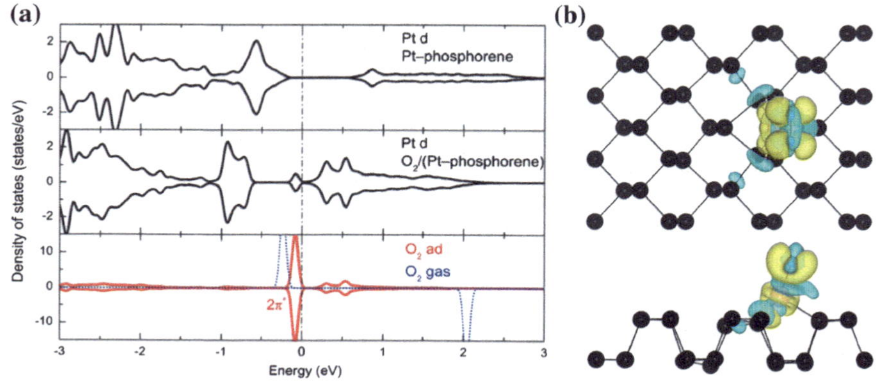

Fig. 8 Local density of states (LDOS) in Pt-phosphorene and O_2-(Pt-phosphorene) systems (**a**), and charge density difference in the O_2-(Pt-phosphorene) system (**b**), in which the yellow region (+0.002 e/Å^3) and the cyan region (−0.002 e/Å^3) denote the increase and loss of the electron density, respectively [98]

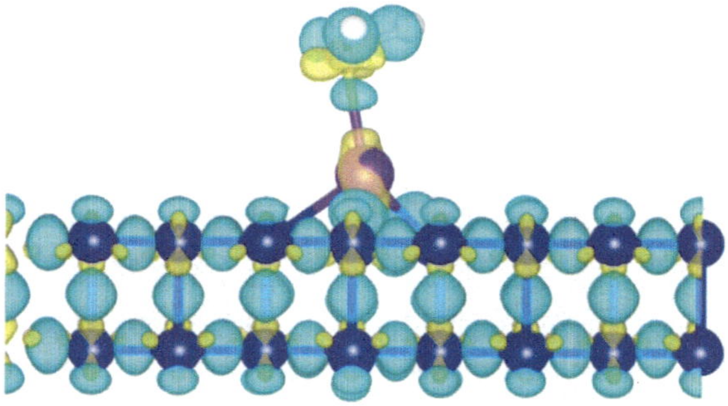

Fig. 9 Charge difference density plot for the H_2S/Pt@BP system; the yellow and cyan colors show the charge accumulation and depletion areas, respectively [100]

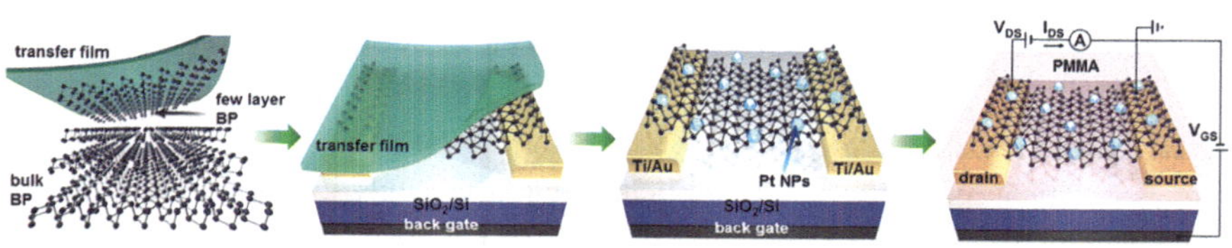

Fig. 10 Schematic presentation of the steps for fabrication of Pt-decorated BP hydrogen sensors via dry transfer [101]

with an adsorption energy of −1.284 eV [100]. Pt-decorated BP could be successfully applied to H_2 sensing with excellent reproducibility and high sensitivity at room temperature (Fig. 10) [17, 101]. Pt-decorated phosphorene also helps with the capture of NO_2 and SO_2 molecules [90]. Ni-decorated phosphorene showed improved stability at ambient conditions and was successful as a catalyst in the semi-hydrogenation of phenylacetylene to styrene [102].

BP decoration with Cu, Ag, and Au from group IB gives rise to n-type impurity states with no substantial changes in the band gap of the semiconductor [83–86]. They can be successful for use as potential electrocatalysts in electrolytic cells [99]. Particularly, Cu has smaller electronegativity and, therefore, denotes electron to the BP surface, thus upshifting the Fermi level toward the conduction band and effectively tuning its work function [24, 83]. Au adatom induces a stable magnetism (0.96 μ_B,) in BP to make it a spin-gapless semiconductor [86, 98]. A tunable highly stable and low-noise response for NO_2 is attained after the decoration of phosphorene with noble metals such as Au [17]. Decoration of BP-NH_2 nanosheets with Au nanoparticles proved efficient in the catalytic reduction of 4-nitrophenol [20]. Au-decorated phosphorene is also considered as a potential surface-supported catalyst [88]. Decoration with Ag could increase the near-infrared (NIR) absorption of BP dramatically [103]. BP decoration with Zn [82] and Hg^{2+} [18] from group IIB has also been reported for adsorption and FET applications.

Decoration with Al, Ga, and In from group III makes the BP monolayer n-type, offering low effective mass of electrons (~0.4 m_e) and thus better mobilities [84, 86, 88, 96, 97, 104]. Nonmetallic adatoms such as B could not strengthen the adsorption of H_2 and CO_2 [93, 95]. Thermal stability of BP was enhanced drastically with Al decoration [13]. HCN can be strongly captured by the low-price and environmentally friendly Al-decorated BP system with a large adsorption energy of −2.166 eV [100]. Even stronger adsorption events have been predicted for acidic gases (CO_2, NO_2, and SO_2) [90]. On the contrary, BP decoration with C, Si, Ge, and Sn from group IVA was found to produce mid-gap states, which are not favored for an effective modification of BP for electronic devices and chemical sensors [84, 93, 95].

None of the N, P, and As elements from group VA can strictly change the BP semiconductor type, and their adsorptive properties are almost negligible [84, 93, 95]. Some authors predicted, however, that the N adatom can result in a p-type semiconductor [86]. DFT studies have shown that a P adatom can stabilize in a B position [81]. The same inertness as mentioned above is also true for the nonmetallic O, S, and Se elements from group VIA [86, 93, 95]. Decoration with tellurium boosts the transport properties and ambient stability of BP FET devices, however [105]. Meanwhile, O prefers a T site for adsorption on phosphorene [13].

While none of the F, Cl, and Br from group VIIA could make the BP monolayer either n-type or p-type, they could shift the Fermi level to the valence band maximum (VBM), which might help increase the BP hole density [84]. Upon increasing the size of halogen adsorbed BP, the absorption coefficient declines while moving toward the visible region [89]. The band gap and the effective mass of the charge carrier can be regulated by the concentration of the adatom [106]. BP fluorination leads to improved long-term stability thanks to the remarkable antioxidation and antihydration properties of the electronegative F adatom. Hence, F-decorated BP showed a promising air-stable photothermal behavior for more than a week [27]. Data from first-principles calculations have shown that a halogen-intercalated BP bilayer structure can serve as an ideal multiferroic nonvolatile memory (NVM) device with mobile magnetism that can be controlled by an external electric field, which in turn renders electrical writing and magnetic reading [107].

5 Black Phosphorus Doping

The progress in the substitutional doping of BP is summarized in the following. This section is structured in terms of different groups of elements according to the periodic table of elements for a more convenient follow-up.

Although the substitutional doping of H has been claimed to be experimentally viable, it is not expected that the H-dopant induces any new magnetism in the monolayer BP owing to the pairing of the valence electrons of dopant with those from the adjacent P atoms [108]. The alkali atoms from the IA group (Li and Na) donate their electrons in the outer *s*-orbitals through ionic bonds with P. Although causing little structural changes [109], the alkali doping makes, the band gap decrease by pushing down the conduction band of BP, implying that the electronic and optical properties could be adjusted [110, 111]. These metal-doped BP monolayers were significantly more reactive than a pristine phosphorene [109, 110, 112]. For instance, the electron–phonon coupling at the doping level of 1/8 of monolayer has been so strong that has eventually led to a superconducting phase [111].

Group IIA dopants are beneficial in sensing applications. Phosphorene behaves p-type and becomes more stable upon doping with calcium. NH_3 binds strongly to the Ca-doped BP structures along the armchair direction [94]. Ca-doped black phosphorene was also beneficial in sensing and removing the tetrachlorodibenzo-*p*-dioxin (TCDD) molecule [113]. Doping of BP with Sc from group IIIB was thermodynamically more favorable in divacancy (DV) complexes than in those having a monovacancy (MV). The defect–transition metal complexes maintained their intrinsic semiconducting properties with no spin-polarized state, but the doping also induced a local magnetic moment with remarkable spin-flip and exchange-splitting energies, thus making them applicable to the spintronic devices (Fig. 11) [114–117].

Theoretical findings of both band structure and density of states have indicated that the BP monolayers doped with most of the lanthanide series of elements (La, Ce, Pr, Nd, Pm, Eu, Gd) except the Pm-doped case involve no substantial changes in the original semiconducting behavior, giving band gaps close to that of the pristine phosphorene (about 0.90 eV). Interestingly, the Eu-doped structure hole doping, which produces a band crossing the Fermi level. The present study suggests that Ln-doped phosphorene can be used as a potential next-generation dilute magnetic semiconductor [118, 119].

The presence of divacancies has also been essential for BP doping with Ti from group IVB [114]. Ti-BP with one P vacancy presents dilute magnetic semiconductor (DMS) properties [116] but rather formed with high energy [115]. Interestingly, the energy gap with spin polarization was observed for the substitutional Ti doping [120], in which non-bonding *d* states are partially filled, thus creating large and localized

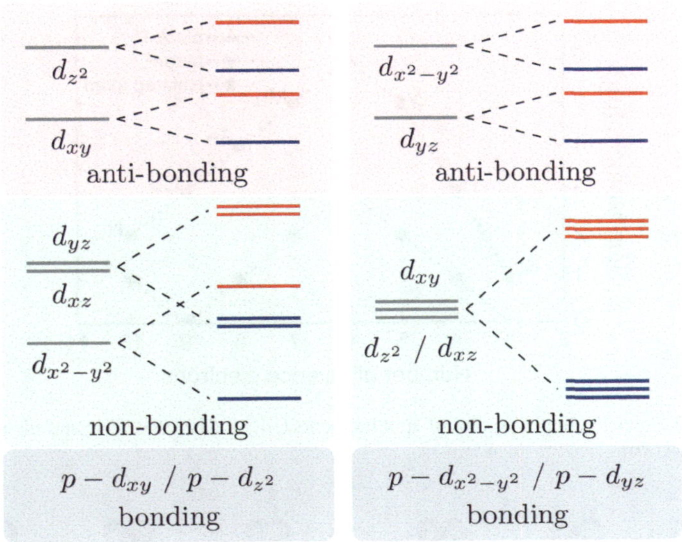

Fig. 11 Simplified presentation of the observed trends in the magnetic moment in the TM doped at MV (left panel) and DV (right panel), which involve the distribution of electrons in the non-binding states, thus generating high-spin solutions [114]

spin moments [116]. This behavior can be effectively modulated by the strain [117]. Ti-doped phosphorene is a good detector for TCDD [113]. BP doping with V (group VB) is more favored at the divacancy defects [114]. This doping induces magnetic moments [115, 117] with DMS properties [116], and leads to a half-metallic state according to calculations with the generalized gradient approximation (GGA) and that incorporating the Hubbard U term (GGA-U) [120].

A similar discussion as those given above is valid for the BP doping with Cr from group VIB and Mn from group VIIB, in which the charge transfer and intra-atomic charge redistribution in the transition metal (TM) atoms led to the magnetic moments of 3 and 2 μ_B, respectively [114–117, 120]. A small biaxial strain induced a magnetic transition in Mn-doped phosphorene [117]. Overall, Cr showed the highest spin moment among the 3*d* TM series of dopants (Fig. 12) [116].

The trends of changes with the group VIIIB (Fe, Co, Ni) dopants were more or less similar [114, 121]. A strong affinity for doping was observed, and the charge transfer and binding energy both expectedly correlated with the electronegativity of the dopant ($\gamma_{Fe} < \gamma_{Co} < \gamma_{Ni}$) [110]. The Co-doped monolayer showed no magnetic state, however [115]. The amount of magnetic moment of each TM dopant was suppressed relative to its bulk value [115]. Nevertheless, one may note the half-metallic state in the Fe-doped BP monolayer [115]. Doping with Fe and Ni induces magnetic moments to phosphorene (Fig. 13) [115]. The Fe dopant is an active site for the catalytic reduction of nitrogen to ammonia [122]. For a Co magnetic impurity, bonding states were wholly empty or filled, thus producing a nonmagnetic material

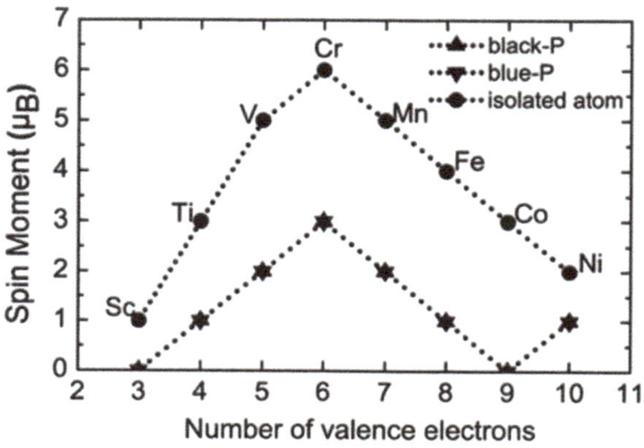

Fig. 12 The Slater-Pauling-type plot of spin moments (μ_B) for isolated TMs and their doping in phosphorene [116]

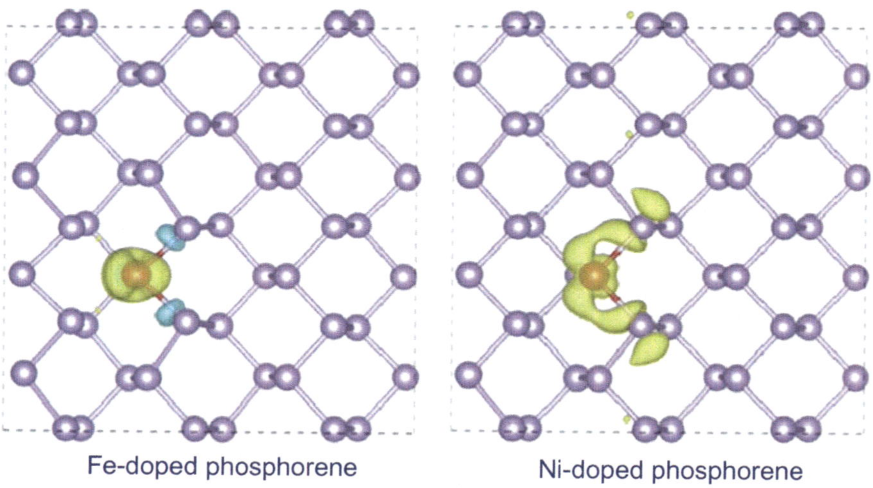

Fig. 13 Isosurface plots of the spin-charge density for Fe and Ni doped BP monolayer (isosurface values set up to ± 0.003 e/Å^3); yellow and blue colors illustrate spin-up and spin-down charge densities, respectively [117]

[116]. However, it became possible to achieve a ferromagnetic-to-antiferromagnetic phase transition in the *p*-doped BP monolayers by including holes or electrons with the aid of a potential gate [123]. Co-doping enhances the sensitivity of BP as a gas sensor of CO with an adsorption energy of 1.31 eV [18].

Phosphorus substitution with the group VIIIB noble metals (Pd and Pt) led to a strong metal–substrate interaction in the modified BP monolayers. Additionally, the mid-gap states due to the dopant suggest new pathways for the charge carrier excitation and recombination [110]. DFT calculations have indicated low formation energies (high adsorption energies) upon doping with Pt [124]. Such doped BP structures adsorbed the NO and CO molecules slightly stronger than the pristine phosphorene [112, 124]. Black phosphorene doping with the IB (Cu, Ag, Au) elements has the same general effects as those mentioned previously [114]. For instance, no spin-polarized state was found in the case of the Cu-doped phosphorene system, which retained its semiconducting band gap after doping [115, 120]. Au doping caused phosphorene to undergo a magnetic-to-nonmagnetic transition [112]. Significant charge transfer values were observed during the interactions of such doped materials with the NO and CO molecules [124]. Zn (from group IIB) interacts either very weakly with black phosphorene or does not bind at all owing to its closed-shell electronic configuration [114].

Substitutional dopants with an odd number of valence electrons, e.g. group IIIA (B, Al and Ga) are known to preserve a semiconducting feature [125]. The low concentration of boron impurity, however, turns black phosphorene from a semiconductor to a semimetal with increased chemical reactivity [126]. Such doped systems are expected to be thermodynamically stable [108, 125]. Doping of B could not induce magnetism to black phosphorene owing to the strong electron pairing and hybridization of *sp* orbitals between the *n*-type dopant and the adjacent P atoms [108, 125, 127]. Similarly, the aluminum impurity did not provoke a spin-polarized state [128].

Meanwhile, the anisotropic nature of the transport properties was reported to decline upon Al and Ga substitution [129]. Phosphorene doping with B and the consequential high reactivity showed promising for potential metal-free SO_2 catalysis, but with adsorption that was too strong for sensing application [130]. B-doped phosphorene was also applied successfully as a frustrated Lewis pair catalyst for hydrogenation of small ketones, nitriles, and olefins [131]. Boron doping was not desired for CH_3OH sensing [132]. However, the B-doped system demonstrated an indirect bandgap and strong anisotropy for optical devices [133]. A co-doping of BP monolayer with B and V led to a significantly distorted lattice with a smaller band gap relative to the primitive structure [134]. Al-doping remarkably improved the electron mobility of the *n*-type BP transistor [25]. The built-in electric field effectively separated NIR photogenerated electron–hole pairs with no requirement of an external bias, thus leading to a promising photovoltaic performance (an open-circuit voltage responsivity of ca. 15.7×10^3 VW^{-1}) at room temperature [25]. The Al-doped phosphorene demonstrated an indirect-to-direct bandgap transition upon CO adsorption [112].

Doping with C, Si, and Ge from group IVA, all with even number of valence electrons, not only bestowed metallic properties to black phosphorene, but also led to highly anisotropic transport properties [125, 127, 129, 135]. Low-concentration doping of BP nanoribbons with C and Si and edge passivation led to a low-bias negative differential resistance (NDR) behavior with values in the order of 10^5–10^8 for the peak-to-valley ratio (PVR) [135]. Such doping increases the energy gap of BP nanoribbons [136] but is not proper for methanol sensing [132]. Some authors point to the observation that the doping sites relax toward a planar geometry for C and

Si doping. This, in turn, gives rise to an acceptor π state in the upper band gap, far from the valence band, thus ruling out the existence of any doping effect [137]. Ab initio total energy calculations suggested that the substitutional carbon dopant forms chemical bonds with three neighboring phosphorus atoms at the grain boundary defects through sp^2-like hybridization, leaving one valence electron unpaired [138]. Spin-polarized computations have revealed that the C doping induces magnetism and tailors the band structure of black phosphorene [138]. Interestingly, the magnetic state was observed in the Si-doped BP monolayer system [128]. Si-doped phosphorene can also strongly adsorb the SO_2 gaseous molecules [130].

The isovalence doping with N and As from group VA preserves the semiconducting characteristics and anisotropic transport properties of black phosphorene [125, 129]. Therefore, no magnetism is induced owing to the pairing of the valence electrons of dopant with those from the neighboring P atoms [108]. Another DFT study revealed that nitrogen doping generates –N–P–P–P–N– chains, which further transform into –P–N–P–N– phosphorene lines with increasing the N concentration. These events are accompanied by the decreased chemical reactivity of N-doped phosphorene [126]. Even though the perfect lattice of BP monolayer could be a work function sensor of alcohol molecules [139], the N doping could slightly improve the adsorption capability of black phosphorene toward methanol [132].

The O, S, and Se doping (from group VIA) not only grants metallic properties to BP but also provides highly anisotropic transport properties [125, 128, 129]. These impurities adopt the 8-N coordination rule as normally found in amorphous semiconducting materials, rather than the simpler substitution geometries in tetrahedral semiconductors [137]. Spin-polarized calculations have suggested that the O doping effectively tunes the band structure and generates magnetism in black phosphorene [138]. Whereas the ground states of Si, O, and Se doped systems were magnetic and half-metallic owing to the generation of a nonbonding $3p$ electron of the adjacent P atom, the charged dopants S^+ and Se^+ induced no magnetism [108, 127]. The basically anionic species, such as O and S present strong binding energy, which indicates that the presence of these impurities in black phosphorene is much likely [110]. The S doping could break the intrinsic space inversion symmetry of the pristine phosphorene, thus leading to a Cs symmetry. A linear circular photogalvanic effect (PGE) is in turn induced in both armchair and zigzag directions [140]. The Se-doped BP crystals have demonstrated high crystallinity and uniformity with highly improved external quantum efficiency (2993%), large on/off current ratio (10^5), and high hole mobility (561 $cm^2V^{-1}s^{-1}$) at ambient conditions [28].

Doping with F, Cl, Br, and I from group VIIA led to a large structure deformation at the doping site of black phosphorene [108]. Cl-doping led to the appearance of a magnetic state in the doped BP monolayer [128]. According to another study, however, such substitutional doping can induce no magnetism in the doped structure [108]. DFT studies have revealed that F acts as a p-type dopant [127].

References

1. Carvalho, A., Wang, M., Zhu, X., Rodin, A.S., Su, H., Castro Neto, A.H.: Phosphorene: from theory to applications. Nat. Rev. Mater. **1**, 16061 (2016). https://doi.org/10.1038/natrevmats.2016.61
2. Du, Y., Luo, Z., Liu, H., Xu, X., Ye, P.D.: Anisotropic properties of black phosphorus. In: Avouris, P., Low, T., Heinz, T.F. (eds.) 2D Materials: Properties and Devices, pp. 413–434. Cambridge University Press, Cambridge (2017). https://doi.org/10.1017/9781316681619.023
3. Jing, Y., Zhang, X., Zhou, Z.: Phosphorene: what can we know from computations? WIREs Comput. Mol. Sci. **6**(1), 5–19 (2016). https://doi.org/10.1002/wcms.1234
4. Ghashghaee, M., Ghambarian, M.: Adsorption of toxic mercury, lead, cadmium, and arsenic ions on black phosphorous nanosheet: first-principles calculations. Struct. Chem. **30**(1), 85–96 (2019). https://doi.org/10.1007/s11224-018-1173-6
5. Greenwood, N.N., Earnshaw, A.: Chemistry of the Elements. Elsevier Science (2012)
6. Guo, H., Lu, N., Dai, J., Wu, X., Zeng, X.C.: Phosphorene nanoribbons, phosphorus nanotubes, and van der Waals multilayers. J. Phys. Chem. C **118**(25), 14051–14059 (2014). https://doi.org/10.1021/jp505257g
7. Bridgman, P.W.: Two new modifications of phosphorus. J. Am. Chem. Soc. **36**(7), 1344–1363 (1914). https://doi.org/10.1021/ja02184a002
8. Lei, W., Liu, G., Zhang, J., Liu, M.: Black phosphorus nanostructures: recent advances in hybridization, doping and functionalization. Chem. Soc. Rev. **46**(12), 3492–3509 (2017). https://doi.org/10.1039/c7cs00021a
9. Keyes, R.W.: The electrical properties of black phosphorus. Phys. Rev. **92**(3), 580–584 (1953). https://doi.org/10.1103/PhysRev.92.580
10. Liu, H., Neal, A.T., Zhu, Z., Luo, Z., Xu, X., Tománek, D., Ye, P.D.: Phosphorene: an unexplored 2D semiconductor with a high hole mobility. ACS Nano **8**(4), 4033–4041 (2014). https://doi.org/10.1021/nn501226z
11. Favron, A., Gaufrès, E., Fossard, F., Phaneuf-L'Heureux, A.-L., Tang, N.Y.W., Lévesque, P.L., Loiseau, A., Leonelli, R., Francoeur, S., Martel, R.: Photooxidation and quantum confinement effects in exfoliated black phosphorus. Nat. Mater. **14**, 826 (2015). https://doi.org/10.1038/nmat4299
12. Liu, H., Du, Y., Deng, Y., Ye, P.D.: Semiconducting black phosphorus: synthesis, transport properties and electronic applications. Chem. Soc. Rev. **44**(9), 2732–2743 (2015). https://doi.org/10.1039/c4cs00257a
13. Feng, X., Kulish, V.V., Wu, P., Liu, X., Ang, K.-W.: Anomalously enhanced thermal stability of phosphorene via metal adatom doping: an experimental and first-principles study. Nano Res. **9**(9), 2687–2695 (2016). https://doi.org/10.1007/s12274-016-1156-0
14. Wood, J.D., Wells, S.A., Jariwala, D., Chen, K.-S., Cho, E., Sangwan, V.K., Liu, X., Lauhon, L.J., Marks, T.J., Hersam, M.C.: Effective passivation of exfoliated black phosphorus transistors against ambient degradation. Nano Lett. **14**(12), 6964–6970 (2014). https://doi.org/10.1021/nl5032293
15. Zhu, H., McDonnell, S., Qin, X., Azcatl, A., Cheng, L., Addou, R., Kim, J., Ye, P.D., Wallace, R.M.: Al_2O_3 on black phosphorus by atomic layer deposition: an in situ interface study. ACS Appl. Mater. Interfaces **7**(23), 13038–13043 (2015). https://doi.org/10.1021/acsami.5b03192
16. Chen, Y., Ren, R., Pu, H., Chang, J., Mao, S., Chen, J.: Field-effect transistor biosensors with two-dimensional black phosphorus nanosheets. Biosens. Bioelectron. **89**, 505–510 (2017). https://doi.org/10.1016/j.bios.2016.03.059
17. Cho, S.-Y., Koh, H.-J., Yoo, H.-W., Jung, H.-T.: Tunable chemical sensing performance of black phosphorus by controlled functionalization with noble metals. Chem. Mater. **29**(17), 7197–7205 (2017). https://doi.org/10.1021/acs.chemmater.7b01353
18. Guo, Z., Chen, S., Wang, Z., Yang, Z., Liu, F., Xu, Y., Wang, J., Yi, Y., Zhang, H., Liao, L., Chu, P.K., Yu, X.-F.: Metal-ion-modified black phosphorus with enhanced stability and transistor performance. Adv. Mater. **29**(42), 1703811 (2017). https://doi.org/10.1002/adma.201703811

19. Zhao, Y., Wang, H., Huang, H., Xiao, Q., Xu, Y., Guo, Z., Xie, H., Shao, J., Sun, Z., Han, W., Yu, X.-F., Li, P., Chu, P.K.: Surface coordination of black phosphorus for robust air and water stability. Angew. Chem. Int. Ed. **55**(16), 5003–5007 (2016). https://doi.org/10.1002/anie.201512038
20. Zhang, L., Gao, L.-F., Li, L., Hu, C.-X., Yang, Q.-Q., Zhu, Z.-Y., Peng, R., Wang, Q., Peng, Y., Jin, J., Zhang, H.-L.: Negatively charged 2D black phosphorus for highly efficient covalent functionalization. Mater. Chem. Front. **2**(9), 1700–1706 (2018). https://doi.org/10.1039/c8qm00237a
21. Liu, Y., Gao, P., Zhang, T., Zhu, X., Zhang, M., Chen, M., Du, P., Wang, G.-W., Ji, H., Yang, J., Yang, S.: Azide passivation of black phosphorus nanosheets: covalent functionalization affords ambient stability enhancement. Angew. Chem. Int. Ed. **58**(5), 1479–1483 (2019). https://doi.org/10.1002/anie.201813218
22. van Druenen, M., Davitt, F., Collins, T., Glynn, C., O'Dwyer, C., Holmes, J.D., Collins, G.: Covalent functionalization of few-layer black phosphorus using iodonium salts and comparison to diazonium modified black phosphorus. Chem. Mater. **30**(14), 4667–4674 (2018). https://doi.org/10.1021/acs.chemmater.8b01306
23. Sun, Z., Xie, H., Tang, S., Yu, X.-F., Guo, Z., Shao, J., Zhang, H., Huang, H., Wang, H., Chu, P.K.: Ultrasmall black phosphorus quantum dots: synthesis and use as photothermal agents. Angew. Chem. **127**(39), 11688–11692 (2015). https://doi.org/10.1002/ange.201506154
24. Koenig, S.P., Doganov, R.A., Seixas, L., Carvalho, A., Tan, J.Y., Watanabe, K., Taniguchi, T., Yakovlev, N., Castro Neto, A.H., Özyilmaz, B.: Electron doping of ultrathin black phosphorus with Cu adatoms. Nano Lett. **16**(4), 2145–2151 (2016). https://doi.org/10.1021/acs.nanolett.5b03278
25. Liu, Y., Cai, Y., Zhang, G., Zhang, Y.-W., Ang, K.-W.: Al-doped black phosphorus p–n homojunction diode for high performance photovoltaic. Adv. Funct. Mater. **27**(7), 1604638 (2017). https://doi.org/10.1002/adfm.201604638
26. Xiang, D., Han, C., Wu, J., Zhong, S., Liu, Y., Lin, J., Zhang, X.-A., Ping, HuW, Özyilmaz, B., Neto, A.H.C., Wee, A.T.S., Chen, W.: Surface transfer doping induced effective modulation on ambipolar characteristics of few-layer black phosphorus. Nat. Commun. **6**, 6485 (2015). https://doi.org/10.1038/ncomms7485
27. Tang, X., Liang, W., Zhao, J., Li, Z., Qiu, M., Fan, T., Luo, C.S., Zhou, Y., Li, Y., Guo, Z., Fan, D., Zhang, H.: Fluorinated phosphorene: electrochemical synthesis, atomistic fluorination, and enhanced stability. Small **13**(47), 1702739 (2017). https://doi.org/10.1002/smll.201702739
28. Xu, Y., Yuan, J., Fei, L., Wang, X., Bao, Q., Wang, Y., Zhang, K., Zhang, Y.: Selenium-doped black phosphorus for high-responsivity 2D photodetectors. Small **12**(36), 5000–5007 (2016). https://doi.org/10.1002/smll.201600692
29. Shao, J., Xie, H., Huang, H., Li, Z., Sun, Z., Xu, Y., Xiao, Q., Yu, X.-F., Zhao, Y., Zhang, H., Wang, H., Chu, P.K.: Biodegradable black phosphorus-based nanospheres for in vivo photothermal cancer therapy. Nat. Commun. **7**, 12967 (2016). https://doi.org/10.1038/ncomms12967, https://www.nature.com/articles/ncomms12967#supplementary-information
30. Zhang, Y., Sun, W., Luo, Z.-Z., Zheng, Y., Yu, Z., Zhang, D., Yang, J., Tan, H.T., Zhu, J., Wang, X., Yan, Q., Dou, S.X.: Functionalized few-layer black phosphorus with superwettability towards enhanced reaction kinetics for rechargeable batteries. Nano Energy **40**, 576–586 (2017). https://doi.org/10.1016/j.nanoen.2017.09.002
31. Zhu, X., Zhang, T., Sun, Z., Chen, H., Guan, J., Chen, X., Ji, H., Du, P., Yang, S.: Black phosphorus revisited: a missing metal-free elemental photocatalyst for visible light hydrogen evolution. Adv. Mater. **29**(17), 1605776 (2017). https://doi.org/10.1002/adma.201605776
32. Shao, L., Sun, H., Miao, L., Chen, X., Han, M., Sun, J., Liu, S., Li, L., Cheng, F., Chen, J.: Facile preparation of NH_2-functionalized black phosphorene for the electrocatalytic hydrogen evolution reaction. J. Mater. Chem. A **6**(6), 2494–2499 (2018). https://doi.org/10.1039/c7ta10884b
33. Zhu, X., Zhang, T., Jiang, D., Duan, H., Sun, Z., Zhang, M., Jin, H., Guan, R., Liu, Y., Chen, M., Ji, H., Du, P., Yan, W., Wei, S., Lu, Y., Yang, S.: Stabilizing black phosphorus nanosheets via edge-selective bonding of sacrificial C60 molecules. Nat. Commun. **9**(1), 4177 (2018). https://doi.org/10.1038/s41467-018-06437-1

34. Xu, G.-L., Chen, Z., Zhong, G.-M., Liu, Y., Yang, Y., Ma, T., Ren, Y., Zuo, X., Wu, X.-H., Zhang, X., Amine, K.: Nanostructured black phosphorus/ketjenblack–multiwalled carbon nanotubes composite as high performance anode material for sodium-ion batteries. Nano Lett. **16**(6), 3955–3965 (2016). https://doi.org/10.1021/acs.nanolett.6b01777
35. Doganov, R.A., O'Farrell, E.C.T., Koenig, S.P., Yeo, Y., Ziletti, A., Carvalho, A., Campbell, D.K., Coker, D.F., Watanabe, K., Taniguchi, T., Neto, A.H.C., Özyilmaz, B.: Transport properties of pristine few-layer black phosphorus by van der Waals passivation in an inert atmosphere. Nat. Commun. **6**, 6647 (2015). https://doi.org/10.1038/ncomms7647
36. Du, Y., Yang, L., Zhou, H., Ye, P.D.: Performance enhancement of black phosphorus field-effect transistors by chemical doping. IEEE Electron Device Lett. **37**(4), 429–432 (2016). https://doi.org/10.1109/led.2016.2535905
37. Mu, H., Lin, S., Wang, Z., Xiao, S., Li, P., Chen, Y., Zhang, H., Bao, H., Lau, S.P., Pan, C., Fan, D., Bao, Q.: Black phosphorus-polymer composites for pulsed lasers. Adv. Opt. Mater. **3**(10), 1447–1453 (2015). https://doi.org/10.1002/adom.201500336
38. Kasap, S., Capper, P.: Springer Handbook of Electronic and Photonic Materials. Springer International Publishing (2017)
39. Sugahara, S., Takamura, Y., Shuto, Y., Yamamoto, S.: Field-effect spin-transistors. In: Xu, Y., Awschalom, D.D., Nitta, J. (eds.) Handbook of Spintronics, pp. 1243–1279. Springer Netherlands, Dordrecht (2016). https://doi.org/10.1007/978-94-007-6892-5_44
40. Eswaraiah, V., Zeng, Q., Long, Y., Liu, Z.: Black phosphorus nanosheets: synthesis, characterization and applications. Small **12**(26), 3480–3502 (2016). https://doi.org/10.1002/smll.201600032
41. Sorkin, V., Cai, Y., Ong, Z., Zhang, G., Zhang, Y.W.: Recent advances in the study of phosphorene and its nanostructures. Crit. Rev. Solid State Mater. Sci. **42**(1), 1–82 (2017). https://doi.org/10.1080/10408436.2016.1182469
42. Akhtar, M., Anderson, G., Zhao, R., Alruqi, A., Mroczkowska, J.E., Sumanasekera, G., Jasinski, J.B.: Recent advances in synthesis, properties, and applications of phosphorene. npj 2D Mater. Appl. **1**(1), 5 (2017). https://doi.org/10.1038/s41699-017-0007-5
43. Qiao, H., Wei, Q.: Functional nanofibers in lithium-ion batteries. In: Wei, Q. (ed.) Functional Nanofibers and their Applications, pp. 197–208. Woodhead Publishing (2012). https://doi.org/10.1533/9780857095640.2.197
44. Khandelwal, A., Mani, K., Karigerasi, M.H., Lahiri, I.: Phosphorene—the two-dimensional black phosphorous: properties, synthesis and applications. Mater. Sci. Eng.: B **221**(Supplement C), 17–34 (2017). https://doi.org/10.1016/j.mseb.2017.03.011
45. McCann, J., Bryson, D.: Smart Clothes and Wearable Technology. Elsevier Science (2009)
46. Kakaei, K., Esrafili, M.D., Ehsani, A.: Introduction to catalysis. In: Kakaei, K., Esrafili, M.D., Ehsani, A. (eds.) Interface Science and Technology, vol. 27, pp. 1–21. Elsevier (2019). https://doi.org/10.1016/B978-0-12-814523-4.00001-0
47. Makhlouf, A.S.H., Tiginyanu, I.: Nanocoatings and Ultra-Thin Films: Technologies and Applications. Elsevier Science (2011)
48. Peixoto, A.C., Silva, A.F.: Smart devices: micro- and nanosensors. In: Rodrigues, L., Mota, M. (eds.) Bioinspired Materials for Medical Applications, pp. 297–329. Woodhead Publishing (2017). https://doi.org/10.1016/B978-0-08-100741-9.00011-5
49. Choi, J.R., Yong, K.W., Choi, J.Y., Nilghaz, A., Lin, Y., Xu, J., Lu, X.: Black phosphorus and its biomedical applications. Theranostics **8**(4), 1005–1026 (2018). https://doi.org/10.7150/thno.22573
50. İşmal, Ö.E., Paul, R.: Composite textiles in high-performance apparel. In: McLoughlin, J., Sabir, T. (eds.) High-Performance Apparel, pp. 377–420. Woodhead Publishing (2018). https://doi.org/10.1016/B978-0-08-100904-8.00019-5
51. Sawant, S.N.: Development of biosensors from biopolymer composites. In: Sadasivuni, K.K., Ponnamma, D., Kim, J., Cabibihan, J.J., AlMaadeed, M.A. (eds.) Biopolymer composites in electronics, pp. 353–383. Elsevier (2017). https://doi.org/10.1016/B978-0-12-809261-3.00013-9

52. Bhagyaraj, S.M., Oluwafemi, O.S., Kalarikkal, N., Thomas, S.: Applications of Nanomaterials: Advances and Key Technologies. Elsevier Science (2018)
53. Jothi, L., Nageswaran, G.: Plasma modified polymeric materials for biosensors/biodevice applications. In: Thomas, S., Mozetič, M., Cvelbar, U., Špatenka, P., Praveen, K.M. (eds.) Non-Thermal Plasma Technology for Polymeric Materials, pp. 409–437. Elsevier (2019). https://doi.org/10.1016/B978-0-12-813152-7.00015-9
54. VanEngelenburg, S.B., Palmer, A.E.: Fluorescent biosensors of protein function. Curr. Opin. Chem. Biol. **12**(1), 60–65 (2008). https://doi.org/10.1016/j.cbpa.2008.01.020
55. Hirsch, A., Hauke, F.: Post-Graphene 2D chemistry: the emerging field of molybdenum disulfide and black phosphorus functionalization. Angew. Chem. Int. Ed. **57**(16), 4338–4354 (2018). https://doi.org/10.1002/anie.201708211
56. Ling, X., Wang, H., Huang, S., Xia, F., Dresselhaus, M.S.: The renaissance of black phosphorus. Proc. Natl. Acad. Sci. **112**(15), 4523 (2015). https://doi.org/10.1073/pnas.1416581112
57. Ryder, C.R., Wood, J.D., Wells, S.A., Hersam, M.C.: Chemically tailoring semiconducting two-dimensional transition metal dichalcogenides and black phosphorus. ACS Nano **10**(4), 3900–3917 (2016). https://doi.org/10.1021/acsnano.6b01091
58. Ryder, C.R., Wood, J.D., Wells, S.A., Yang, Y., Jariwala, D., Marks, T.J., Schatz, G.C., Hersam, M.C.: Covalent functionalization and passivation of exfoliated black phosphorus via aryl diazonium chemistry. Nat. Chem. **8**, 597 (2016). https://doi.org/10.1038/nchem.2505
59. Boukhvalov, D.W., Rudenko, A.N., Prishchenko, D.A., Mazurenko, V.G., Katsnelson, M.I.: Chemical modifications and stability of phosphorene with impurities: a first principles study. Phys. Chem. Chem. Phys. **17**(23), 15209–15217 (2015). https://doi.org/10.1039/c5cp01901j
60. Drissi, L.B., Sadki, S., Sadki, K.: Half-oxidized phosphorene: band gap and elastic properties modulation. J. Phys.: Condens. Matter **28**(14), 145501 (2016). https://doi.org/10.1088/0953-8984/28/14/145501
61. Wang, G., Pandey, R., Karna, S.P.: Phosphorene oxide: stability and electronic properties of a novel two-dimensional material. Nanoscale **7**(2), 524–531 (2015). https://doi.org/10.1039/c4nr05384b
62. Dai, J., Zeng, X.C.: Structure and stability of two dimensional phosphorene with =O or =NH functionalization. RSC Adv. **4**(89), 48017–48021 (2014). https://doi.org/10.1039/c4ra02850c
63. Kistanov, A.A., Cai, Y., Zhou, K., Dmitriev, S.V., Zhang, Y.-W.: The role of H_2O and O_2 molecules and phosphorus vacancies in the structure instability of phosphorene. 2D Mater. **4**(1), 015010 (2016). https://doi.org/10.1088/2053-1583/4/1/015010
64. Cai, Y., Ke, Q., Zhang, G., Zhang, Y.-W.: Energetics, charge transfer, and magnetism of small molecules physisorbed on phosphorene. J. Phys. Chem. C **119**(6), 3102–3110 (2015). https://doi.org/10.1021/jp510863p
65. Sadki, S., Drissi, L.B.: Tunable optical and excitonic properties of phosphorene via oxidation. J. Phys.: Condens. Matter **30**(25), 255703 (2018). https://doi.org/10.1088/1361-648x/aac403
66. Çakır, D., Sahin, H., Peeters, F.M.: Tuning of the electronic and optical properties of single-layer black phosphorus by strain. Phys. Rev. B **90**(20), 205421 (2014)
67. Ienco, A., Manca, G., Peruzzini, M., Mealli, C.: Modelling strategies for the covalent functionalization of 2D phosphorene. Dalton Trans. **47**(48), 17243–17256 (2018). https://doi.org/10.1039/c8dt03628d
68. Scotognella, F., Kriegel, I., Sassolini, S.: Covalent functionalized black phosphorus quantum dots. Opt. Mater. **75**, 521–524 (2018). https://doi.org/10.1016/j.optmat.2017.11.016
69. Sofer, Z., Luxa, J., Bouša, D., Sedmidubský, D., Lazar, P., Hartman, T., Hardtdegen, H., Pumera, M.: The covalent functionalization of layered black phosphorus. Angew. Chem. Int. Ed. **56**(33), 9891–9896 (2017). https://doi.org/10.1002/anie.201705722
70. Li, Q., Zhou, Q., Niu, X., Zhao, Y., Chen, Q., Wang, J.: Covalent functionalization of black phosphorus from first-principles. J Phys. Chem. Lett. **7**(22), 4540–4546 (2016). https://doi.org/10.1021/acs.jpclett.6b02192
71. Abellán, G., Lloret, V., Mundloch, U., Marcia, M., Neiss, C., Görling, A., Varela, M., Hauke, F., Hirsch, A.: Noncovalent functionalization of black phosphorus. Angew. Chem. Int. Ed. **55**(47), 14557–14562 (2016). https://doi.org/10.1002/anie.201604784

72. Lee, M., Park, Y.H., Kang, E.B., Chae, A., Choi, Y., Jo, S., Kim, Y.J., Park, S.-J., Min, B., An, T.K., Lee, J., In, S.-I., Kim, S.Y., Park, S.Y., In, I.: Highly efficient visible blue-emitting black phosphorus quantum dot: mussel-inspired surface functionalization for bioapplications. ACS Omega **2**(10), 7096–7105 (2017). https://doi.org/10.1021/acsomega.7b01058
73. Li, Y., Liu, Z., Hou, Y., Yang, G., Fei, X., Zhao, H., Guo, Y., Su, C., Wang, Z., Zhong, H., Zhuang, Z., Guo, Z.: Multifunctional nanoplatform based on black phosphorus quantum dots for bioimaging and photodynamic/photothermal synergistic cancer therapy. ACS Appl. Mater. Interfaces **9**(30), 25098–25106 (2017). https://doi.org/10.1021/acsami.7b05824
74. Hu, H., Gao, H., Gao, L., Li, F., Xu, N., Long, X., Hu, Y., Jin, J., Ma, J.: Covalent functionalization of black phosphorus nanoflakes by carbon free radicals for durable air and water stability. Nanoscale **10**(13), 5834–5839 (2018). https://doi.org/10.1039/c7nr06085h
75. Artel, V., Guo, Q., Cohen, H., Gasper, R., Ramasubramaniam, A., Xia, F., Naveh, D.: Protective molecular passivation of black phosphorus. npj 2D Mater. Appl. **1**(1), 6 (2017). https://doi.org/10.1038/s41699-017-0004-8
76. Feng, Q., Liu, H., Zhu, M., Shang, J., Liu, D., Cui, X., Shen, D., Kou, L., Mao, D., Zheng, J., Li, C., Zhang, J., Xu, H., Zhao, J.: Electrostatic functionalization and passivation of water-exfoliated few-layer black phosphorus by poly dimethyldiallyl ammonium chloride and its ultrafast laser application. ACS Appl. Mater. Interfaces **10**(11), 9679–9687 (2018). https://doi.org/10.1021/acsami.8b00556
77. Yu, X., Zhang, S., Zeng, H., Wang, Q.J.: Lateral black phosphorene P-N junctions formed via chemical doping for high performance near-infrared photodetector. Nano Energy **25**, 34–41 (2016). https://doi.org/10.1016/j.nanoen.2016.04.030
78. Zhang, R., Li, B., Yang, J.: A first-principles study on electron donor and acceptor molecules adsorbed on phosphorene. J. Phys. Chem. C **119**(5), 2871–2878 (2015). https://doi.org/10.1021/jp5116564
79. Jing, Y., Tang, Q., He, P., Zhou, Z., Shen, P.: Small molecules make big differences: molecular doping effects on electronic and optical properties of phosphorene. Nanotechnology **26**(9), 095201 (2015). https://doi.org/10.1088/0957-4484/26/9/095201
80. Cao, Y., Tian, X., Gu, J., Liu, B., Zhang, B., Song, S., Fan, F., Chen, Y.: Covalent functionalization of black phosphorus with conjugated polymer for information storage. Angew. Chem. Int. Ed. **57**(17), 4543–4548 (2018). https://doi.org/10.1002/anie.201712675
81. Kulish, V.V., Malyi, O.I., Persson, C., Wu, P.: Adsorption of metal adatoms on single-layer phosphorene. Phys. Chem. Chem. Phys. **17**(2), 992–1000 (2015). https://doi.org/10.1039/c4cp03890h
82. Lei, S.Y., Yu, Z.Y., Shen, H.Y., Sun, X.L., Wan, N., Yu, H.: CO adsorption on metal-decorated phosphorene. ACS Omega **3**(4), 3957–3965 (2018). https://doi.org/10.1021/acsomega.8b00133
83. Sun, H., Shang, Y., Yang, Y., Guo, M.: Realization of N-Type semiconducting of phosphorene through surface metal doping and work function study. J. Nanomater. **2018**, 9 (2018). https://doi.org/10.1155/2018/6863890
84. Rastogi, P., Kumar, S., Bhowmick, S., Agarwal, A., Chauhan, Y.S.: Effective doping of mono-layer phosphorene by surface adsorption of atoms for electronic and spintronic applications. IETE J. Res. **63**(2), 205–215 (2017). https://doi.org/10.1080/03772063.2016.1243020
85. Zhu, H., Zhou, C., Wang, X., Chen, X., Yang, W., Wu, Y., Lin, W.: Doping behaviors of adatoms adsorbed on phosphorene. Phys. Status Solidi B **253**(6), 1156–1166 (2016). https://doi.org/10.1002/pssb.201552586
86. Ding, Y., Wang, Y.: Structural, electronic, and magnetic properties of adatom adsorptions on black and blue phosphorene: a first-principles Study. J. Phys. Chem. C **119**(19), 10610–10622 (2015). https://doi.org/10.1021/jp5114152
87. Zhiyuan, Y., Shuangying, L., Neng, W., Shan, L., Haiyun, S., Hong, Y.: Effect of metal adatoms on hydrogen adsorption properties of phosphorene. Mater. Res. Express **4**(4), 045503 (2017). https://doi.org/10.1088/2053-1591/aa6ac0
88. Hu, T., Hong, J.: First-principles study of metal adatom adsorption on black phosphorene. J. Phys. Chem. C **119**(15), 8199–8207 (2015). https://doi.org/10.1021/acs.jpcc.5b01300

89. Musle, V., Choudhary, S.: Tuning the optical properties of phosphorene by adsorption of alkali metals and halogens. Opt. Quant. Electron. **50**(7), 285 (2018). https://doi.org/10.1007/s11082-018-1548-3
90. Kuang, A., Kuang, M., Yuan, H., Wang, G., Chen, H., Yang, X.: Acidic gases (CO_2, NO_2 and SO_2) capture and dissociation on metal decorated phosphorene. Appl. Surf. Sci. **410**(Supplement C), 505–512 (2017). https://doi.org/10.1016/j.apsusc.2017.03.135
91. Huang, G.Q., Xing, Z.W., Xing, D.Y.: Prediction of superconductivity in Li-intercalated bilayer phosphorene. Appl. Phys. Lett. **106**(11), 113107 (2015). https://doi.org/10.1063/1.4916100
92. Han, C., Hu, Z., Gomes, L.C., Bao, Y., Carvalho, A., Tan, S.J.R., Lei, B., Xiang, D., Wu, J., Qi, D., Wang, L., Huo, F., Huang, W., Loh, K.P., Chen, W.: Surface functionalization of black phosphorus via potassium toward high-performance complementary devices. Nano Lett. **17**(7), 4122–4129 (2017). https://doi.org/10.1021/acs.nanolett.7b00903
93. Zhang, H.-p., Hu, W., Du, A., Lu, X., Zhang, Y.-p., Zhou, J., Lin, X., Tang, Y.: Doped phosphorene for hydrogen capture: a DFT study. Appl. Surf. Sci. **433**(Supplement C), 249–255 (2018). https://doi.org/10.1016/j.apsusc.2017.09.243
94. Lalitha, M., Nataraj, Y., Lakshmipathi, S.: Calcium decorated and doped phosphorene for gas adsorption. Appl. Surf. Sci. **377**(Supplement C), 311–323 (2016). https://doi.org/10.1016/j.apsusc.2016.03.119
95. Zhang, H.-p., Du, A., Shi, Q.-b., Zhou, Y., Zhang, Y., Tang, Y.: Adsorption behavior of CO_2 on pristine and doped phosphorenes: a dispersion corrected DFT study. J. CO_2 Utilization **24**, 463–470 (2018). https://doi.org/10.1016/j.jcou.2018.02.005
96. Pan, Y., Wang, Y., Ye, M., Quhe, R., Zhong, H., Song, Z., Peng, X., Yu, D., Yang, J., Shi, J., Lu, J.: Monolayer phosphorene-metal contacts. Chem. Mater. **28**(7), 2100–2109 (2016). https://doi.org/10.1021/acs.chemmater.5b04899
97. Zhang, X., Pan, Y., Ye, M., Quhe, R., Wang, Y., Guo, Y., Zhang, H., Dan, Y., Song, Z., Li, J., Yang, J., Guo, W., Lu, J.: Three-layer phosphorene-metal interfaces. Nano Res. **11**(2), 707–721 (2018). https://doi.org/10.1007/s12274-017-1680-6
98. Luo, Y., Ren, C., Wang, S., Li, S., Zhang, P., Yu, J., Sun, M., Sun, Z., Tang, W.: Adsorption of transition metals on black phosphorene: a first-principles study. Nanoscale Res. Lett. **13**(1), 282 (2018). https://doi.org/10.1186/s11671-018-2696-x
99. Zhang, Y., Dong, N., Tao, H., Yan, C., Huang, J., Liu, T., Robertson, A.W., Texter, J., Wang, J., Sun, Z.: Exfoliation of stable 2D black phosphorus for device fabrication. Chem. Mater. **29**(15), 6445–6456 (2017). https://doi.org/10.1021/acs.chemmater.7b01991
100. Kuang, A., Ran, Y., Peng, B., Kuang, M., Wang, G., Yuan, H., Tian, C., Chen, H.: Adsorption and decomposition of metal decorated phosphorene toward H_2S, HCN and NH_3 molecules. Appl. Surf. Sci. **473**, 242–250 (2019). https://doi.org/10.1016/j.apsusc.2018.12.131
101. Lee, G., Jung, S., Jang, S., Kim, J.: Platinum-functionalized black phosphorus hydrogen sensors. Appl. Phys. Lett. **110**(24), 242103 (2017). https://doi.org/10.1063/1.4985708
102. Caporali, M., Serrano-Ruiz, M., Telesio, F., Heun, S., Nicotra, G., Spinella, C., Peruzzini, M.: Decoration of exfoliated black phosphorus with nickel nanoparticles and its application in catalysis. Chem. Commun. **53**(79), 10946–10949 (2017). https://doi.org/10.1039/c7cc05906j
103. Hu, Z., Li, Q., Lei, B., Wu, J., Zhou, Q., Gu, C., Wen, X., Wang, J., Liu, Y., Li, S., Zheng, Y., Lu, J., He, J., Wang, L., Xiong, Q., Wang, J., Chen, W.: Abnormal near-infrared absorption in 2D black phosphorus induced by Ag nanoclusters surface functionalization. Adv. Mater. **30**(43), 1801931 (2018). https://doi.org/10.1002/adma.201801931
104. Prakash, A., Cai, Y., Zhang, G., Zhang, Y.-W., Ang, K.-W.: Black phosphorus N-Type field-effect transistor with ultrahigh electron mobility via aluminum adatoms doping. Small **13**(5), 1602909 (2017). https://doi.org/10.1002/smll.201602909
105. Yang, B., Wan, B., Zhou, Q., Wang, Y., Hu, W., Lv, W., Chen, Q., Zeng, Z., Wen, F., Xiang, J., Yuan, S., Wang, J., Zhang, B., Wang, W., Zhang, J., Xu, B., Zhao, Z., Tian, Y., Liu, Z.: Te-doped black phosphorus field-effect transistors. Adv. Mater. **28**(42), 9408–9415 (2016). https://doi.org/10.1002/adma.201603723

106. Nahas, S., Ghosh, B., Bhowmick, S., Agarwal, A.: First-principles cluster expansion study of functionalization of black phosphorene via fluorination and oxidation. Phys. Rev. B **93**(16), 165413 (2016). https://doi.org/10.1103/PhysRevB.93.165413
107. Yang, Q., Xiong, W., Zhu, L., Gao, G., Wu, M.: Chemically functionalized phosphorene: two-dimensional multiferroics with vertical polarization and mobile magnetism. J. Am. Chem. Soc. **139**(33), 11506–11512 (2017). https://doi.org/10.1021/jacs.7b04422
108. Zheng, H., Zhang, J., Yang, B., Du, X., Yan, Y.: A first-principles study on the magnetic properties of nonmetal atom doped phosphorene monolayers. Phys. Chem. Chem. Phys. **17**(25), 16341–16350 (2015). https://doi.org/10.1039/c5cp00916b
109. Ehlen, N., Senkovskiy, B.V., Fedorov, A.V., Perucchi, A., Di Pietro, P., Sanna, A., Profeta, G., Petaccia, L., Grüneis, A.: Evolution of electronic structure of few-layer phosphorene from angle-resolved photoemission spectroscopy of black phosphorous. Phys. Rev. B **94**(24), 245410 (2016). https://doi.org/10.1103/PhysRevB.94.245410
110. Suvansinpan, N., Hussain, F., Zhang, G., Chiu, C.H., Cai, Y., Zhang, Y.-W.: Substitutionally doped phosphorene: electronic properties and gas sensing. Nanotechnology **27**(6), 065708 (2016). https://doi.org/10.1088/0957-4484/27/6/065708
111. Sanna, A., Fedorov, A.V., Verbitskiy, N.I., Fink, J., Krellner, C., Petaccia, L., Chikina, A., Usachov, D.Y., Grüneis, A., Profeta, G.: First-principles and angle-resolved photoemission study of lithium doped metallic black phosphorous. 2D Materials **3**(2), 025031 (2016). https://doi.org/10.1088/2053-1583/3/2/025031
112. Sun, X., Luan, S., Shen, H., Lei, S.: Effect of metal doping on carbon monoxide adsorption on phosphorene: a first-principles study. Superlattices Microstruct. **124**, 168–175 (2018). https://doi.org/10.1016/j.spmi.2018.09.037
113. Zhang, H-p, Hou, J-l, Wang, Y., Tang, P-p, Zhang, Y-p, Lin, X-y, Liu, C., Tang, Y.: Adsorption behavior of 2, 3, 7, 8-tetrachlorodibenzo-p-dioxin on pristine and doped black phosphorene: a DFT study. Chemosphere **185**, 509–517 (2017). https://doi.org/10.1016/j.chemosphere.2017.06.120
114. Babar, R., Kabir, M.: Transition metal and vacancy defect complexes in phosphorene: a spintronic perspective. J. Phys. Chem. C **120**(27), 14991–15000 (2016). https://doi.org/10.1021/acs.jpcc.6b05069
115. Wang, Y., Pham, A., Li, S., Yi, J.: Electronic and magnetic properties of transition-metal-doped monolayer black phosphorus by defect engineering. J. Phys. Chem. C **120**(18), 9773–9779 (2016). https://doi.org/10.1021/acs.jpcc.6b00981
116. Yu, W., Zhu, Z., Niu, C.-Y., Li, C., Cho, J.-H., Jia, Y.: Dilute magnetic semiconductor and half-metal behaviors in 3*d* transition-metal doped black and blue phosphorenes: a first-principles study. Nanoscale Res. Lett. **11**(1), 77 (2016). https://doi.org/10.1186/s11671-016-1296-x
117. Zhai, C., Dai, X., Li, W., Ma, Y., Wang, T., Tang, Y.: Strain tuning of magnetism in transition-metal atom doped phosphorene. Superlattices Microstruct. **101**(Supplement C), 49–56 (2017). https://doi.org/10.1016/j.spmi.2016.10.090
118. Yuan, Z., Li, N.: Manipulating the magnetic moment in phosphorene by lanthanide atom doping: a first-principle study. RSC Adv. **6**(94), 92048–92056 (2016). https://doi.org/10.1039/c6ra14546a
119. Luan, Z., Zhao, L., Chang, H., Sun, D., Tan, C., Huang, Y.: First-principles study on electronic structures and magnetic properties of Eu-doped phosphorene. Superlattices Microstruct. **111**(Supplement C), 816–823 (2017). https://doi.org/10.1016/j.spmi.2017.07.039
120. Hashmi, A., Hong, J.: Transition metal doped phosphorene: first-principles study. J. Phys. Chem. C **119**(17), 9198–9204 (2015). https://doi.org/10.1021/jp511574n
121. Arabieh, M., Azar, Y.T.: In silico insight into ammonia adsorption on pristine and X-doped phosphorene (X=B, C, N, O, Si, and Ni). Appl. Surf. Sci. **396**(Supplement C), 1411–1419 (2017). https://doi.org/10.1016/j.apsusc.2016.11.175
122. Wei, Z., Zhang, Y., Wang, S., Wang, C., Ma, J.: Fe-doped phosphorene for the nitrogen reduction reaction. J. Mater. Chem. A **6**(28), 13790–13796 (2018). https://doi.org/10.1039/c8ta03989e

123. Seixas, L., Carvalho, A., Castro Neto, A.H.: Atomically thin dilute magnetism in Co-doped phosphorene. Phys. Rev. B **91**(15), 155138 (2015). https://doi.org/10.1103/PhysRevB.91.155138
124. Qi-Hang, Y., Yong, J., Wei, Z., Bo-Zhao, W., Jiu-Ren, Y., Ping, Z., Yan-Huai, D.: Noble metal atoms doped phosphorene: electronic properties and gas adsorption ability. Mater. Res. Express **4**(4), 045703 (2017)
125. Yu, W., Zhu, Z., Niu, C.-Y., Li, C., Cho, J.-H., Jia, Y.: Anomalous doping effect in black phosphorene using first-principles calculations. Phys. Chem. Chem. Phys. **17**(25), 16351–16358 (2015). https://doi.org/10.1039/c5cp01732g
126. Boukhvalov, D.W.: The atomic and electronic structure of nitrogen- and boron-doped phosphorene. Phys. Chem. Chem. Phys. **17**(40), 27210–27216 (2015). https://doi.org/10.1039/c5cp05071e
127. Yang, L., Mi, W., Wang, X.: Tailoring magnetism of black phosphorene doped with B, C, N, O, F, S and Se atom: a DFT calculation. J. Alloys Compounds **662**(Supplement C), 528–533 (2016). https://doi.org/10.1016/j.jallcom.2015.12.095
128. Khan, I., Hong, J.: Manipulation of magnetic state in phosphorene layer by non-magnetic impurity doping. New J. Phys. **17**(2), 023056 (2015). https://doi.org/10.1088/1367-2630/17/2/023056
129. Guo, C., Xia, C., Fang, L., Wang, T., Liu, Y.: Tuning anisotropic electronic transport properties of phosphorene via substitutional doping. Phys. Chem. Chem. Phys. **18**(37), 25869–25878 (2016). https://doi.org/10.1039/c6cp04508a
130. Yang, Q., Meng, R.S., Jiang, J.K., Liang, Q.H., Tan, C.J., Cai, M., Sun, X., Yang, D.G., Ren, T.L., Chen, X.P.: First-principles study of sulfur dioxide sensor based on phosphorenes. IEEE Electron Dev. Lett. **37**(5), 660–662 (2016). https://doi.org/10.1109/led.2016.2543243
131. Zhao, J., Liu, X., Chen, Z.: Frustrated lewis pair catalysts in two dimensions: B/Al-doped phosphorenes as promising catalysts for hydrogenation of small unsaturated molecules. ACS Catal. **7**(1), 766–771 (2017). https://doi.org/10.1021/acscatal.6b02727
132. Liu, D., Shi, Y., Tao, L., Yan, D., Chen, R., Wang, S.: First-principles study of methanol adsorption on heteroatom-doped phosphorene. Chin. Chem. Lett. **30**(1), 207–210 (2019). https://doi.org/10.1016/j.cclet.2018.01.041
133. Wu, Z.-F., Gao, P.-F., Guo, L., Kang, J., Fang, D.-Q., Zhang, Y., Xia, M.-G., Zhang, S.-L., Wen, Y.-H.: Robust indirect band gap and anisotropy of optical absorption in B-doped phosphorene. Phys. Chem. Chem. Phys. **19**(47), 31796–31803 (2017). https://doi.org/10.1039/c7cp05404a
134. Kong, L.-J., Liu, G.-H., Zhang, Y.-J.: Tuning the electronic and optical properties of phosphorene by transition-metal and nonmetallic atom co-doping. RSC Adv. **6**(13), 10919–10929 (2016). https://doi.org/10.1039/c5ra22004a
135. Guo, C., Wang, T., Xia, C., Liu, Y.: Modulation of electronic transport properties in armchair phosphorene nanoribbons by doping and edge passivation. Sci. Rep. **7**(1), 12799 (2017). https://doi.org/10.1038/s41598-017-13212-7
136. Nasria, A.H.A.: Electronic properties of phosphorene nanoribbons doped with boron, aluminum and carbon. J. Chem. Pharm. Sci. **10**(2), 935–939 (2017)
137. Guo, Y., Robertson, J.: Vacancy and doping states in monolayer and bulk black phosphorus. Sci. Rep. **5**, 14165 (2015). https://doi.org/10.1038/srep14165
138. Zhu, Z.-L., Yu, W.-Y., Ren, X.-Y., Sun, Q., Jia, Y.: Grain boundary in phosphorene and its unique roles on C and O doping. EPL (Europhys. Lett.) **109**(4), 47003 (2015). https://doi.org/10.1209/0295-5075/109/47003
139. Ghambarian, M., Ghashghaee, M., Azizi, Z., Balar, M.: Molecular interactions of MeOH and EtOH with black phosphorus monolayer: a periodic density functional study. Phys. Chem. Res. **7**(2), 435–447 (2019). https://doi.org/10.22036/pcr.2019.172026.1594
140. Xie, Y., Zhang, L., Zhu, Y., Liu, L., Guo, H.: Photogalvanic effect in monolayer black phosphorus. Nanotechnology **26**(45), 455202 (2015). https://doi.org/10.1088/0957-4484/26/45/455202

Black Phosphorous Based Nanodevices

J. Ashtami, S. S. Athira, V. G. Reshma and P. V. Mohanan

Abstract Black phosphorus (BP) has come into sight as a hopeful two dimensional (2D) material from its time of invention in 2014 through flourishing exfoliation method. Devices possessing nonspecific properties in at least one dimension are generally considered as nanodevices. The far reaching attention on BP based nanodevices came from its unique structural, compositional and functional features. Researchers have focused on 2D structures of BP because of the electron distribution and wide band gap peculiarities. On an overall point view, the material is considered to be apt for a plethora of medical as well as non-medical application scenarios. Nevertheless, its applicability is getting hindered by certain characteristic drawbacks. For the reason that BP is relatively unstable in air and aqueous environment, several functionalization strategies have been adopted in recent years. This chapter addresses mainly the different types of BP based nanodevices, its non-medical and medical applications, safety aspects of the material along with assured challenges it possess in application scenarios. The section lends a hand to the readers to grab information on fascinating potentials of BP nanodevices in various applications.

Keywords Black phosphorous · Nanodevices · Biomedical · Drug delivery · Imaging

1 Introduction

Nanodevices are nanoparticles (NPs) that are created for the purpose of intermingling with cells and tissues thereby carry out various specific tasks. A nanodevice, or nanomachine, is defined as a nanoscale automaton or at least one containing nanosized components [1]. The major proportion of the renowned nanodevices is contributed by the imaging tools. Black phosphorus (BP), a recent discovery like a

All Authors contributed equally.

J. Ashtami · S. S. Athira · V. G. Reshma · P. V. Mohanan (✉)
Toxicology Division, Biomedical Technology Wing, Sree Chitra Tirunal Institute for Medical Sciences and Technology, Thiruvananthapuram 695 012, Kerala, India
e-mail: mohanpv10@gmail.com; mohanpv@sctimst.ac.in

Inamuddin et al. (eds.), *Black Phosphorus*, Engineering Materials,
https://doi.org/10.1007/978-3-030-29555-4_2

promising 2D material, meets far-reaching consideration with its device application because of the superior electronic properties.

In 1914, Bridgman synthesized BP single crystals for the first time through his analysis of conversion of white phosphorus under high pressure [2]. Bulk BP crystals are a semiconductor with a direct band gap value of 0.35 eV and charge carrier mobilities of 10,000 cm^2/Vs [3]. Compared to transition metal dichalcogenides, BP is an impressive outlook for broadband photo detection owing to their small and direct band gap. BP seems to be like graphene by their layered structure with only one element; phosphorus. In BP crystal structure, atoms are powerfully connected in-plane appearing as layers, although the layers interact via van der waals forces. Phosphorus atoms in its layered structure have a valence shell configuration of $3s^2 3p^3$ with five valence electrons. Each phosphorus atom undergoes hybridization to form sp^3 hybridized orbitals inorder to facilitate bonding with other atoms. The hybridized orbitals form covalent bonds with four adjoining phosphorus atoms resulting in a puckered structure [4].

The crystal structure of BP is very much influenced by the pressure. Orthorhombic crystalline structure is the largely known stable form of phosphorus at 1 bar pressure. When the pressure reaches 10–11 GPa, BP alters its structure to simple cubic form [5]. Along with the crystalline structure of the BP, other properties like band gap or exciton binding energy, optical property such as luminescence etc. can be fine-tuned [6, 7].

2 Importance of BP Nanodevices

BP is a promising 2D material in device application. It gathers extensive attention because of the advanced characteristics. Currently researchers are focusing more on applications of BP in electronic and optoelectronics. The thickness dependent tunable band gap also makes it a promising candidate for photonics, especially for near and mid-IR applications. A number of illustrations of BP nanodevices by means of diverse range of capabilities are shown in Fig. 1.

BP is a prospective candidate in next-generation electronics. In electronics, transistor constitutes major proportion of fundamental and key applications in electronics. The general criteria for determining the efficiency of transistor includes high on/off ratio, elevated carrier (holes) mobility and elevated conductivity. BP is employed to be a channel material because it exhibit high hole mobilities (10–1000 cm^2/Vs) and massive on/off ratios of 100–10,000 conditional on flake thickness [8–13]. The particular formulates BP; a better-quality stuff for FET while correlating with other 2D materials such as graphene or TMDCs. Despite of the outstanding carrier mobility, graphene cannot achieve low off state current as a consequence of its metallic behaviour. Ambipolar field-effect of BP is a benefit in the manufacture of more multifaceted nanodevices like PN junctions. MoS_2 and WS_2 exhibit noticeable unipolar n-type activities [14]. Photo voltaic devices having BP as the major substrate depends on piling of two nanosheets forming a PN junction [15]. BP nanosheets (BPNS) are

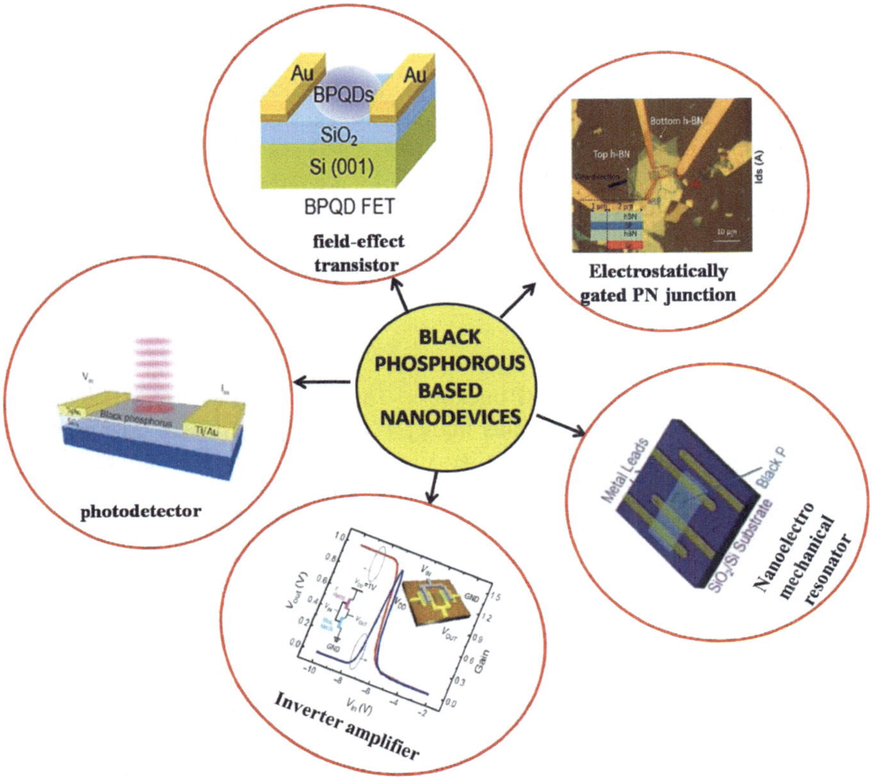

Fig. 1 Categorization of BP based nanodevices

used as nanoelectro-mechanical resonators. It's low mass due to the reduced dimensions make them attractive candidate for the application in resonators [16]. Resembling other 2D semiconducting materials, BP can also be utilized to fabricate several other electronic utensils including circuits, detectors etc.

3 Types of BP Nanodevices

Very rapid fabrication of BP nanodevices is possible after gathering the experience from graphene nanodevices production [10–14, 17]. Different types of BP nanodevices are described below.

3.1 Naonoelectronic Devices

BP is considered to be one of the most hopeful candidates in 2D semiconductors for high speed and power efficient nano electronics. Because they have superior properties including high carrier mobility ~1000 cm^2/Vs, sizable current modulation ratios ~10^5, direct band gap within the range of 0.3 eV (bulk) to 2 eV (monolayer) relies on their breadth [18].

3.2 Thin-Film Transistors

Bottom gated BP transistor with BP thickness ~15 nm have high hole carrier mobility ~1560 cm^2/Vs. This device contains highly P type doped Si with 25 nm Al_2O_3 accumulated as the substrate. 50 nm Ti/Au source and drain contact stacks were created using e-beam lithography and e-beam evaporation after exfoliation of few layers of BP flakes on to the substrate. It also shows high on/off ratio. The device point towards the potential of BP for future high speed and low power nano electronics [19].

In ambient environment, BP transistors cannot be well conserved due to its hygroscopic nature which leads to quick oxidation which causes harsh deterioration of the appliance performance. Such situation could be overcome via an effective dielectric encapsulation method. BP effectively covered by 25 nm Al_2O_3/DuPont Teflon-AF stack to protect from oxidation. This helped for withstanding for 79 days of exposure in ambient environment. Flexible electronics devices and circuits possible by dielectric encapsulated air-stable BP thin-film transistors. P atom outlines three covalent bonds individually by means of its nearest neighbour P atoms in sp^3 hybridizations. Puckered structure results in strong in-plane anisotropy along different orientations relating to electrical, mechanical and thermal properties [20].

Electron transport favors the AC direction for the relatively smaller effective mass; in the meantime, ZZ direction exhibits a higher thermal conductively. This relation between high electron and thermal transport makes BP a promising candidate for thermal electronics with strong in plane selectivity. Due to this unique in-plane anisotropic property, an efficient methodology of determining the in-plane orientation of BP samples becomes rather critical in most practical applications [21].

3.3 Ambipolar Transistors and Circuits

BP has great role in designing of device metrics because of their strong current saturation, high mobilities, sizeable on/off ratio and direct bandgap. These properties make BP suitable for elevated pace digital and analog transistors and circuits on bendable agents [22].

3.4 Flexible BP Ambipolar Inverter

The ambipolar carrying features in made-up BP FETs with the pronounced current modulation got effectively attained through inverter device structure. Ambipolar inverters are fabricated using a comparatively wider BP flake ($t = 15$ nm) with favourable lower bandgap of ~0.3 eV [23]. Ambipolar transport characteristic is mainly due to small band gap value which results in moderately lesser Schottky barrier height of the electron injection. For slender films, Schottky barrier is large and with reduced ambipolarity [24, 25].

3.5 Flexible BP Amplifier

Single-transistor amplifiers are among the known circuit units which serve as signal transforming units. High voltage gain achieved by current dissemination of appliance properties [26]. BP transistor can play role of both popular types of electronic configurations.

4 Non Medical Applications

4.1 Gas Sensors

Several researchers explained about the potential of BP gas sensor [27–30]. According to the Density functional theory, Gases, like NO or NO_2 strongly bind with black phosphorus. Current along the BP increases with adsorption of these gases, while NH_3 lessens current. Majority of the black phosphorous supported gas sensors are made-up as FETs. Yan et al. [30] demonstrated the usage of black phosphorous in hydrogen peroxide sensing.

4.2 Water Splitting Photocatalyst

Photocatalyst is a material that speeds up the light mediated reaction. It creates electron-hole pairs by absorbing light depends on their photocatalytic activity. These processes consist of some important steps:

- Upon light absorption, electron-hole pairs are formed
- Photocatalyst becomes excited
- Migration of generated electron-hole pairs

- In thermodynamically favourable condition, electrons and holes decrease and oxidize the adsorbed moieties at the border.

Semiconductors take up photons from irradiated sunlight in the photocatalytic water splitting. These semiconductors take in photons elevated than their band gap energy [4]. BP is a hopeful applicant for photocatalysis because the absorption scope of black phosphorus can be extended [31, 32] . In 2014, Sa et al. proposed that strain manufacturing of BP make possible its relevance in photocatalysis [33].

4.3 Energy Storage Devices

Efficient energy storage is a major requisite and much thought has been concentrated on to developing reliable energy storage devices. The attractive opto-electronic properties like the high carrier mobility, band-gap characteristics etc. mould BP as an excellent option for energy storage devices. BP based nanodevices have been developed as solar cells, batteries and generators. By controlling BP layer thickness, the bandgap can be fine-tuned to facilitate photon adsorption. This property is explored for fabricating BP for energy conversion in photovoltaics. Chen et al. reported the use of BPQDs at the anode side in order to accelerate the hole extraction of the characteristic p-i-n planar perovskite solar cells. BP fabricated hybrid solar cell showed increased photon conversion efficiency from 14.10 of 16.69%. The improved device performance confirms the potential of BP as a reliable material for solar cells [34]. BP is reported to be a dependable electrode for lithium as well as sodium batteries. The difficulty in fabrication and low electrical conductivity hinders its scope as electrode. Hybridization of phosphorous with carbon is an opted strategy to overcome these obstacles. The developed hybrid BP-Carbon electrodes illustrated improved initial capacity in both sodium and lithium half life batteries with a corresponding value of 1300 and 1700 mA h g^{-1}. Crystalline BP nanoparticle-carbon composite electrode put forward better stability. As per the reported results, lithium and sodium batteries showed a steady cyclic performance [35].

Materials capable of converting thermal energy to electrical energy are termed as thermoelectric materials. Nanomaterials are extensively investigated for its potential as thermoelectric generators. Among various nanomaterials, BP emerges as an ideal preference due to its short thermal conductivity, greater electrical conductivity as well as elevated Seebeck coefficient. Zhang et al. [36] reported that according to first-principles calculation, BP showed superior thermoelectric properties as evident from thermoelectric figure of merit value of 1.1. According to the literature, BP display thermoelectric property along the armchair direction which opens up possibility for developing highly efficient thermoelectric generators based on BP [36]. Air instability is one of the major demerits for restraining its thermoelectric applications. Silver nanoparticle- BP heterostructure was developed so as to act as thermoelectric generators with enhanced performance and improved stability. The fabrication of BP with gold nanoparticles helped to increase the electrical conductivity and power factor. The power factor of gold nanoparticle decoration on to BP surface enhanced 2740 times when compared to thermoelectric generator based on pristine BP alone [37].

5 Biomedical Applications of BP Nanodevices

From the period of its first invention through exfoliation in 2014, BP (also known as phosphorene) has turned into an attractive nanomaterial among scientific community owing to the extraordinary characteristics. Unique optical, mechanical as well as electrical functionalities have turned it to be a perfect material to be exploited in various application scenarios [38]. It is being applied in various fields exceptionally because of their advantageous properties over graphene and other graphene based materials. Although studies are flourished with reports on BP based applications in industrial and commercial fields, medial usages is still have to elucidated more. This could be potentially be attributed to certain inherent drawbacks it possesses. One of the major shortcomings includes lack of stability when exposed to air or water environment [39]. Meanwhile several modification strategies have been put forward to achieve elevated steadiness for BP in the mentioned circumstances. For instance, Li et al. in 2015 have developed ionophore-encapsulated BP sensor which could be exploited for the high throughput detection of ions present in the system. This modified BP has shown excellent stability in air along with improved capabilities of multiplex detection of ions present which was far superior to that of other graphene based materials. The simplicity of such ionophore-modified BP material holds tremendous application potentials in various application fields; notably because of their ability to detect heavy metals present which is usually a difficult task to be achieved [40].

Regarding the obstacles observed during application of BP, the fundamental factor lies in the high reactivity of them with oxygen and water under ambient conditions. This ultimately leads to physical and compositional changes. Many efforts have been devoted for finding out the exact mechanism behind such reactivity of BP. Structures having globule like shape have identified on BP facade via atomic force and electronic microscopic visualizations. P-type doping was observed during drawn out air exposure on the periphery of BP flakes primarily due to atmospheric water absorption [27].

Graphene other 2D materials including dichalcogenides possess relatively high electronic perturbations and hence are employed widely for the purpose of biosensing, gas sensing and water sensing. BPNS have been identified to exhibit superior sensing capacity for NO_2 while being an active agent for Field effect transistor (FET) biosensor. The detection limit the same have been observed to be 5 ppb. Appropriate band gap and carrier mobility contributes to its characteristic high throughput capacity of sensing. Besides the leading role in the perspective of gas and water, the unique features contribute to its application in sensing of biomolecules also. For example, puckered surface morphology of BP aids in its specific interaction with proteins. Zhang et al. in 2015 have analyzed the biological response of phosphorene and graphene using a protein called villin headpiece (HP35). Molecular dynamic studies indicated that graphene was more imprudent in generating protein disruptions than phosphorene. Further molecular level analysis using pseudo graphene and pseudo phosphorene have revealed that surface puckering of the compound have contributed to the weakening of phosphorene induced disruption of the protein. Such understanding is seems to be more helpful in adopting strategies for designing and elucidating protein-nanomaterial interactions. Also it helps to have knowledge about the bioapplication potentials of phosphorene [41]. BP based drug delivery system have received much attention nowadays. In particular, drug delivery prospects via the assistance of near-infrared (NIR) was made possible by means of merging BP sheets with Strontium Chloride ($SrCl_2$); a promising oseteogenic material. BP inherently possesses high NIR absorption as well as light-to-heat conversion capabilities. This property makes it a suitable material to be applied in such drug delivery scenarios which comprise bio-compatible and biodegradable polymers as a drug loading vehicle [42].

Recently BP based nanosheets have attracted significant attention among researchers in analytical science. BP nanosheets (BPNS) were employed as gas and vapour sensors in earlier studies. Afterwards scientists evaluated the potential of BPNS for the sensing of cartian analytes including hydrogen peroxide as well as inorganic ions. They were also checked for application as biomarkers of pathological abnormalities in living system. Song et al. in [43] have reported the fascinating applications of 2D naosheets. Precise discrimation of double stranded and single stranded DNA molecules was the centre of attraction of the particular study. Overall study highlighted the higher level sensitivity of BPNS in fulfilling the task.

Zhou et al. [44] in 2018 got inspired from the observations of high throughput sensing behaviour of BPNS and analyzed the potential of sensing miRNA in a given sample. Here the researchers evaluated the fluorescence quenching nature of BPNS inorder to fabricate a novel biosensor. This could be considered as an ignition for miRNA detection study using 2D material like BPNS. For the study, BPNS was synthesized by liquid exfoliation method and the obtained final fluorescent quenching material exhibited an average linear range of 10–1000 nM, with a detection limit of 9.37 nM with 40 min of detection limit. The study also analyzed the type of interaction between BPNM surface and genetic materials with elevated specificity. Other biomolecules like proteins and amino acids was also analyzed for their susceptibility of sensing by BPNM. The study therefore could be considered as a path breaking discovery of novel biosensor utilizing fluorescence quenching nature of 2D nanomaterial like BPNS.

BP based devices has also been utilized extensively for therapeutic strategies against cancer. Being a material possessing wider band gap, BP displays outstanding tunable properties which could be make use of as a suitable targeting agent for cancer therapeutics. This is achievable for both photothermal therapy (PTT) and photodynamic therapy (PDT). Exfoliated BP monolayers have already been used for the tumor targeting rationales mainly via enhanced permeability and retention effect (EPR effect). Both PTT and PDT rely on the NIR absorbing materials which convert light into heat energy; thereby specifically targeting tumor tissues. This is conceptually correlates with hyperthermia treatment of cancers. The photosensitizing materials used for these approaches should be able to produce reactive oxygen species (ROS) upon induction by NIR which would further turn the tumor tissue to get destroyed by cytotoxic reactions. BP in this regard is a rising star and has attained considerable interest among researchers [45]. Studies have demonstrated that the size of NPs plays a crucial part in tuning the biodistribution pattern of NPs and their fate inside living system. Those nanomaterials used for the tumor targeting purpose should be of the size suitable for pronounced EPR effect. Those with a size in between 20 and 200 nm normally seems to exhibit good retention effect near tumor tissue thereby achieving satisfactory level of destruction. However, NPs below 20 or 10 nm are more likely to get eliminated by phagocytes even within 6 h. This applies in the case of BP quantum dots (BPQDs).

Exceptional blending of biodegradability and biocompatibility of BP based system was analyzed previously by Shao et al. [46]. Poly (lacto-co-glycolic acid) or PLGO polymer encapsulated BPQDs were analyzed for their capabilities in achieving PTT. PLGA served to control the biodegradation rate of BPQDs and also elevated the PTD stability to higher levels. Similar surface functionalization strategies for BP based nanomaterials are mandatory for achieving higher retention effect and also for minimizing deterioration rate at the target sites.

This would also aid BPNPs of varying sizes to retain within tumor tissue for longer time [46]. Another victorious study in this regard was done recently by Qui et al. [47]. The study put emphasis on the development of a BP based hydrogel repository for which they provided a trade name BP@Hydrogel. The major attraction of the study lies in the fact that this hydrogel showed excellent capabilities of degradation when induced by NIR and in this manner releasing the drug of interest (doxorubicin in different concentrations) at the tumor site. BP@Hydrogel was prepared by using low melting point agarose as solidifying agent and PEGylated BPNS.BP@Hydrogel was synthesized by combining PEGylated BPNS and agarose in specified ration. The NIR used for the photothermal effect possessed a wavelength of 808 nm. Moreover it was evident that BPNS exhibited superior photothermal renovation efficiency than BPQDs (38.8% and 28.45% respectively). At tumor site, when the BP@Hydrogel was exposed to NIR, the hydrogel warmed up and due to the photothermal ability of BPNS, it became softer. Further increase in the laser power contributed to the hydrolysis of the ester linkages between the polymer segments. This further expelled out the monomeric residues and then oligomers; finally converted into carbon dioxide and water. Upon degradation of the hydrogel, the encapsulated BPNS were released out. The degradation products obtained from BPNS were phosphate and phosphonate; both of them were non-toxic in the system. The study highlights the degradation nature of the preparation only after treatment; hence ensuring biosafety [47], The functioning principle of BP@Hydrogel during NIR induction for controlled drug release at tumor site is illustrated in Fig. 2.

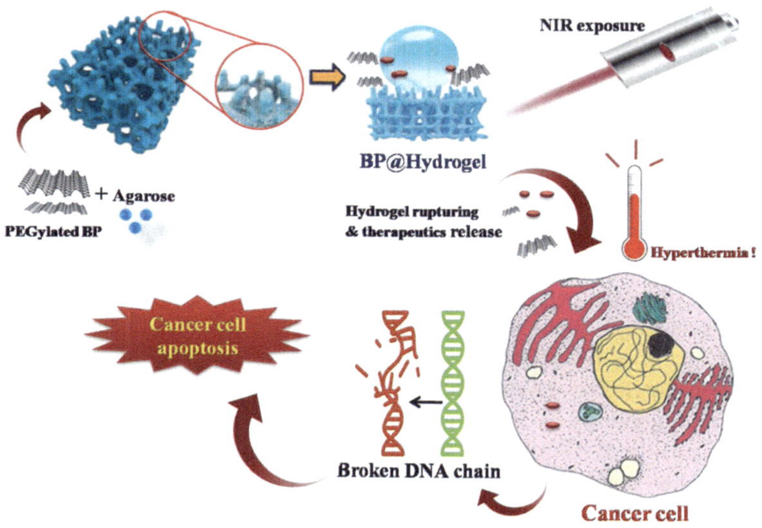

Fig. 2 The functioning principle of BP@Hydrogel during NIR induction for controlled drug release at tumor site

Surface functionalized BP nanomaterials have also attracted thought among researchers in achieving bioimaging more efficiently. Sun et al. in 2016 have demonstrated PEGylated BPNPs which were biocompatible as well as water soluble. One-pot solventless high energy mechanical milling (HEMM) was adopted for the fabrication of the particular stuff. The NPs thus formed showed excellent functionality in photoacoustic imaging (PA) and PTT. Since the approach involved dual functioning of both PA imaging and PTT for cancer treatment, it definitely holds remarkable level of applicability in theranostics. The mentioned study could be regarded as the ignition for the dual purpose of PA and PTT. EPR effect in tumor tissue then accelerated the gathering of water soluble PEGylated BP NPs which further aided PA imaging. The material preferred hepatic and kidney routes for elimination and the animal got returned safely after achieving PTT successfully [48].

Functionalization of monolayers of BPNS by iron oxide NPs was also found exhibit good results in PTT of cancer treatment via NIR induction. These functionalized BP showed ability of producing ROS in a more pronounced manner than bare BPNPs with utmost biocompatibility and stability. It could hopefully serves as the base for Surface Enhanced Raman Scattering (SERS); which further enhance scopes to be applied in the field on biomedicine [49].

5.1 Orthopedic Applications of BP Nanodevices

Various clinical challenges experienced by clinicians all over the world comprise management of distress initiated from imperfections of bone, inflammations and development of cancer and other associated impediments. So far none of the available materials can absolutely meet the requisites for every craniomaxillofacial applications. Therefore surgeons need to have a complete understanding of both advantageous and disadvantageous outcomes the material of interest could generate; and should consider the overall cost, morbidity and likelihood of the patient [50]. Up to now, bone implantation using autogenous source encompasses the finest approach for orthopaedic disease management. Especially because of the high vascularization rate, osseointegration capacity, dislodgement and limited side effects. However at most of the events, it could add some negative consequences like donor site morbidity, additional time requirement for graft harvest, resorption of grafts, and difficulties in molding and limited donor availability for pediatric patients etc. Even though there are numerous alternatives available for overcoming such drawbacks associated, most of them can raise unpredictable consequences and poor osseointegration with nearby bone tissue [51]. The same kind of undesirable reactions are also frequently occurs in allografts and xenografts which would become unable to satisfy the needs.

As discussed earlier, BP assisted delivery of osseoinductive materials like $SrCl_2$ have already came into field with far superior competence. Osteoporosis found associated with post menopausal period have reported to be eradicated with such Sr based drug delivery strategies and thereby reduce vertebral fracture risk factors.

For osteosarcoma therapies, various treatment modalities have been developed based on nanomaterials. Osteosarcoma constitutes a rigorous malignant condition which arises from those primordial cells originated from mesenchymal source inducing demarcation of osteoblast and development of nasty osteoid. Being a revolutionary manufacturing technology, 3D printing embraces incredible potentials to be applied in this field of bone regeneration [52]. Novel synthetic scaffolds have attained enough consideration in bone regeneration facets using sophisticated methods [53]. However fabrication and development of 3D scaffolds for localized cancer therapy is still in the stage infancy. Once interact with biological environment, BPNS tends to react with visible light, oxygen and water; and by this means undergoes disintegration. Degradation products normally found associated with this disintegration are phosphate and phosphonate; which may results in diminishing of photothermal ability. This could ultimately leads to deprived therapeutic efficiency for cancer. Meanwhile there occurs a hidden possibility for applying elemental phosphorus in the likelihood of bone regeneration. However, studies focusing of applying mentioned approach for the treatment of bone regeneration are really scarce. This P can act literally as prompt for osteogenesis and osseointegration. Therefore strategies giving rise to disintegration of BP at the target site could provide elemental phosphorus which inturn stimulate the rate of osteosarcoma therapeutics based on unique physic-chemical properties of BP [46].

Yang et al. in 2018 have made a breakthrough in the field of bone regeneration by employing 2D layers of BPNS. A novel therapeutic strategies combining 2D BPNS with 3D printed bio glass (BG) was fabricated. This bifunctional scaffold was employed for both the treatment of osteosarcoma and subsequent bone regeneration. The scaffold involved a unique combination of photothermal activity contributed by BP and inherent osteogenesis, osteoconduction and osteoinduction roles contributed by 3D BG scaffold. BG induced cell differentiation and proliferation on its surface simultaneously stimulated vascularization and angiogenesis. The predominant role played by scaffold was to act like a matrix to nourish BPNS and to inducing tumor associated tissue defects. The PO_4^{3-} ions released by BPNS obviously accelerated in situ biomineralization in order to promote bone regeneration. The PO_4^{3-} ions served as the ligand for circulating Ca^+ ions which is very essential for the formation of calcium phosphate ($CaPO_4$) during osteogenesis. Hence it is very striking that application of 2D nanomaterials like BPNS can play dual roles like photothermal functioning and promotion of osteogenesis [52].

The aforementioned study emphasized the fact that upon incubation of osteosarcoma cells (Saos-2) with BP/BG scaffolds exhibited excellent growth rate within 3 days of incubation; thereby indicating good biocompatibility. Moreover, those Saos-2 cells grown on BG scaffold alone showed deprived growth rate compared to BP-BG platform. This clearly highlights the information that BP acts like active site for cell proliferation. NIR irradiation caused the death of BP-BG scaffold in pronounced manner than that of those grown on BG alone. This further emphasizes the photothermal activity of BP. The observations from in vitro study ensured feasibility in applying the material for in vivo studies. For analysing the photo thermal efficiency of BP-BG scaffold, 4-weeks old Balb/c nude mice were subjected to Saos-2 cell implantation through subcutaneous route. The animal group administered with BP-BG combination showed relatively increased temperature than the control groups. Localized hyperthermia obtained for BP-BG administered mice was recorded and was found to be increasing from 30 to 55 °C within 1 min and attained 58 °C within 5 min. Whereas, BG alone treated mice did not develop any significant hyperthermia. Furthermore, animals exposed to BP-BG scaffold did not show any significant weight loss; thus ensuring biosafety of the material without any undesirable side effects. Hematoxylin and Eosin (H & E) staining revealed that NIR exposure caused serious necrosis cell lesions in BP-BG treated mice compared to other groups. Complete tumor ablation was obtained this category within 14 days itself. Figure 3 shows an overview of the work and the principle involved.

5.2 *Neurological Applications of BP Nanodevices*

Beyond the possibilities of cancer treatment, BP has been found to possess roles in neuroprotection. Presence of excess levels of transition metals in biological systems can cause serious health consequences which are extremely difficult to eradicate using conventional treatment modalities. Most of the obstacles arise because of the

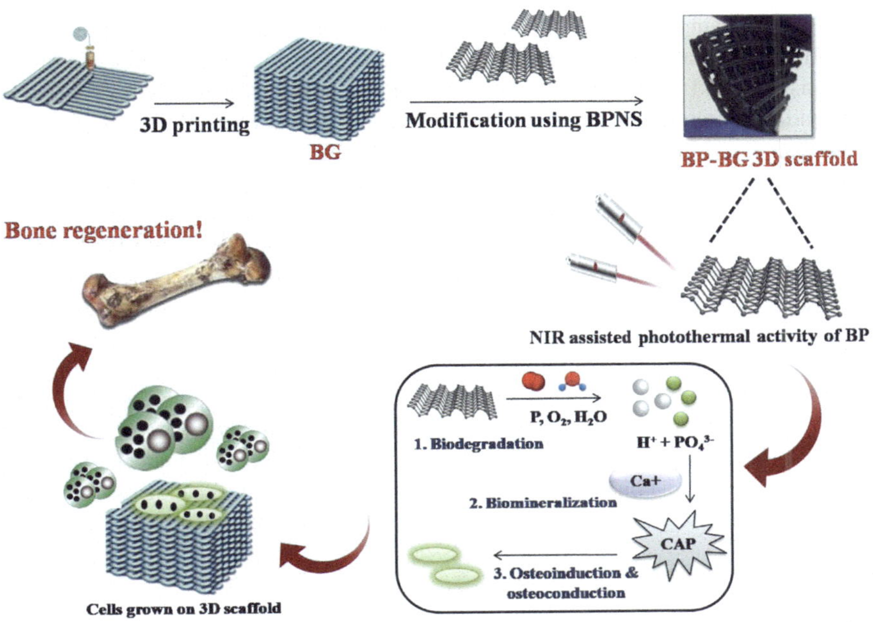

Fig. 3 Working principle of BP-BG 3D scaffold for bone regeneration induced by NIR irradiation

inability of drugs to cross the biological barriers like BBB. Transition metals once invade into brain loci, are more likely to cause the generation of free radicals and hence oxidative stress in neuronal cell community. This would ultimately leads to neuronal cell apoptosis and results in dreadful anomalies in brain including Parkinsonism, Alzheimer's disease etc. Thus chelating agents having property of BBB crossing and scavenging of redox active metal ions is desirable for the treatment of neurodegenerative disorders. Chen et al. in 2017 claimed in a study that BPNS can be regarded as an appropriate material to be applied in such circumstances since they are capable of crossing BBB and capture metal ions wandering in neuronal tissue. Among the various ions tested in cerebrospinal fluid (CSF) such as Ca^{+}, Mg^{2+}, Zn^{2+}, Fe^{2+}, Fe^{3+} and Cu^{2+}, BPNS showed excellent capturing capacity for Cu^{2+}. Total concentration of Cu^{2+} ions in CSF was decreased by 43% whereas rest of the tested ions did not show declining in their amounts. Meanwhile BPNS could protect the neuronal tissue from the ROS related oxidative stress regenerated as a result of Cu^{2+} dishomeostasis.

Elevation in permeability of BPNS in BBB was clearly demonstrated from fitting experiments. NIR mediated BP assault was also studied. Increase in localized hyperthermia was recorded and was found to be only 3.4 °C for BP free solution; served as control. While BP solution of the concentration 20 μg/ml showed an increase in temperature of 23 °C. This draws attention to superior photothermal activity of BP

[54]. BBB crossing pattern of BPNS was studied thereafter using Evans blue staining. Mice head was irradiated using 808 nm laser after intravenous administration of Evans blue and BPNS. Thermal fluctuation was analyzed using thermal camera. The head temperature was unusually increasing and gave a value of 42.2 °C. Control group tested via same route showed only a temperature fluctuation of 35.3 °C. The presence of Evans blue was visible in mice brain alone after 24 h of incubation. Furthermore, there occurred only negligible level of tissue damage as observed from H&E assisted pathological evaluation compared to control group. Inorder to check the long term influence of BP, Balb/c mice was received intravenous injection of BP and irradiated with NIR for 5 min and leave for 7 days. Brain was analyzed after 7 days post exposure using nuclear magnetic resonance imaging and no indications of cerebral thrombosis was observed. Cytotoxicity analysis using different cell lines (HeLa cells, L929 cells, A549 cells and MCF-7 cells) also ensured higher biosafety of BPNS without any cell death. Other tissues including lung, heart, liver, spleen and kidneys were also analyzed for toxicity lesions which also provided negative results; confirming unbeatable biosafety and biocompatibility of BP. Hence the overall study can be exploited for the obliteration of neurodegenerative disorders [54].

5.3 Biosensor Applications

The prime benefit of using BP in opto-electronics is that BP can remarkably act as an upper hand in overcoming the limitations of graphene as well as transition metal dichalcogenides (TMDs) without compromising its advantageous characteristics. Graphene is a highly celebrated 2D material owing to the peculiar physicochemical properties. However, the low ON/OFF current ratio of the allotrope of carbon; graphene, hinders its scope in electronics. TMDs, another potential 2D class of materials and highly researched graphene analogue possess high ON/OFF current ratio but exhibit very low charge mobility. In this scenario, BP step up its position as a reliable option with its appreciable carrier mobility as well as ON/OFF current ratio paving its way for sensing applications.

BP is highly sensitive to electrical perturbation owing to the exceptional carrier mobility. Therefore can sense adsorption or desorption of gaseous molecules based on electrical conductivity measurement. The carrier mobility value of about 1000 $cm^2V^{-1}s^{-1}$ for BP further confirms its potential scope as a biosensor [55]. Sensors with high sensitivity are prerequisite for biomedical diagnosis where the biomarkers may be present in only trace amounts (Fig. 4).

5.4 BP Based Electrochemical Sensor

Electrochemical sensors hold high significance amongst diverse types of biosensors. The electrochemical property of BP was analyzed by Wang et al. by cyclic

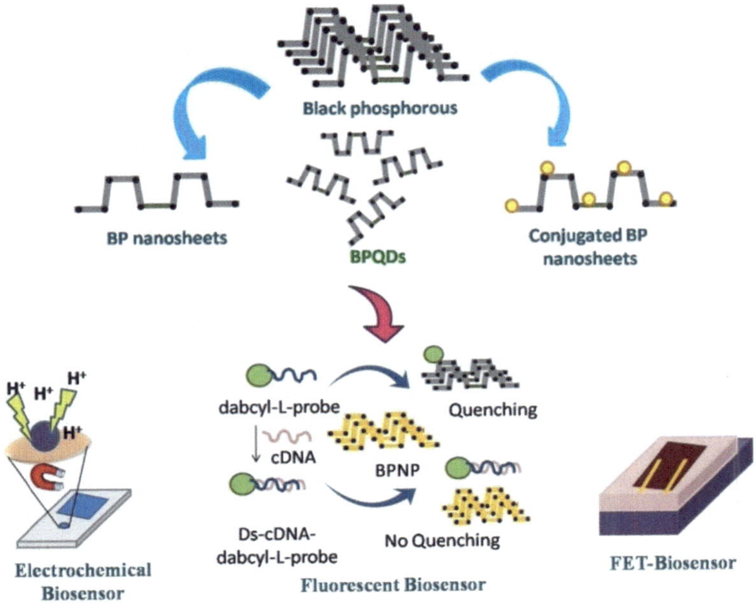

Fig. 4 Some of the BP based biosensors

voltammetric analysis. According to the above mentioned work, the BP crystals synthesized by chemical vapor transport technique was found to oxidize at ~560 mV which indicate its inherent electrochemistry. The result advocates the use of BP in electrochemical sensing [56]. BP nanoparticles prepared by electrochemical exfoliation technique showed potential application in immunoassay as an electrocatalytic tag for the identifying rabbit immunoglobulin G (IgG). The synthesized BP nanoparticles served the purpose of catalyst for hydrogen evolution reaction. Magnetic beads were conjugated with anti-rabbit IgG while BP was used to tag the IgG in samples. Denaturation of protein complex occurs with addition of acid. The hydrogen evolution determines the quantity of IgG. The fabricated biosensor was found to be highly accurate and appreciable precision towards IgG detection. Efficient conductivity as well as the chemical and other electrical properties of BP is good for precise and critical detection of various biomarkers. The low toxicity associated with BP further adds advantage for its sensing application [57].

5.5 *FET Biosensor*

BP based nanodevices as FET biosensor applications is a highly explored field and research findings shows promising performance for biomedical applications.

Chen et al. reported that BP based FET biosensor labelled with gold nanoparticles showed excellent performance in protein detection. In the above mentioned study, the developed BP device could successfully detect human immunoglobulin G (IgG) with utmost sensitivity. The BP nanodevices passivated with Al_2O_3 maintained good stability and suppressed degradation against ambient atmospheric conditions [58]. BP based FET biosensors are excellent option for diagnosis and point of care applications.

Similarly, BP based FET can be successfully used for the detection of harmful chemical toxicants like mercury. Mercury (Hg) is a highly poisonous element and even a trace concentration higher than 1 parts per billion (ppb) can cause harsh health issues including DNA damage, brain damage, birth defects, dysregulation in functioning of kidney, interruption in immune responses etc. Hence, it is vital to confirm that the water, food and air we are exposed is scrutinized for Hg presence. But the present techniques available such as inductively coupled plasma-mass spectroscopy (ICP-MS), inductively coupled plasma- atomic emission spectroscopy (ICP-AES), surface plasmon resonance (SPR) sensing etc. require highly sophisticated system and are time consuming even though the results are reliable. Li et al. reported highly selective BP FET sensors with a sensing limit of 0.01 ppb and time constant of detection as 3 s [59].

5.6 Gas Biosensors

BP possesses high gas sensitivity. This property of BP can be applied for environmental screening to detect toxic gases. BP shows superior sensing performance compared to other 2D material. In a comparative study between graphene, MoS_2 and black phosphorous, BP was found to exhibit higher sensitivity, selectivity, dynamic response. BP showed 20 folds superior sensitivity relative to graphene as well as MoS_2 [60]. The BP based gas biosensors show a thickness dependent sensitivity when explored for chemical sensing of NO_2 molecules. Depending on the BP layer thickness, BP based biosensor showed an extreme sensitivity towards NO_2 with sensitivity striking its highest at 4.8 nm. The fabricated BP based nanodevice showed superior selectivity towards NO_2 in comparison to other gases like carbon monoxide, hydrogen gas, and hydrogen sulphide etc [28]. The mechanism involved in chemical sensing capability of BP based biosensor is based on charge transfer as evident from the Langmuir Isotherm. The Raman spectra analysis of BP based nanodevices confirms appreciable stability even after exposure to NO_2, opening up scope for medical diagnosis [27].

5.7 Fluorescent Biosensors

The optical characteristics and fluorescent property of BP is explored for developing BP based fluorescent biosensors. Most interestingly, the optical properties of BP is tunable that opens up possibility for manipulating the properties so as to impart sensitive detection. Fine tuning the BP particle size could effectively manipulate the optical properties as well as fluorescence of BP. The relatively high quantum yield, tunable absorption-emission characteristics and stability of BP makes them good fluorophores [61]. The efficiency of BP nanoparticles as fluorescent detection agent is demonstrated in work reported by Yew et al. The BP NPs conjugated to single stranded DNA showed high sensitivity towards nucleic acid detection. The results support the possibility for developing BP based biosensor nanodevices for disease diagnosis [62]. Assessment of the activity of acetylcholinesterase is vital in diagnosis, drug test and identification of toxins. BP quantum dots (ODs) with green fluorescence were reported to be efficient fluorescent probe for the sensing the activity of acetylcholinesterase. 5,5′-dithiobis-(2-nitrobenzoic acid) (DTNB) can react with moieties having thiol group resulting in the formation of 2-nitro-5-thiobenzoate (TNB) anion. Therefore, DTNB can be utilized for sensing the presence of thiol groups in molecules like glutathione (GSH)/L-cysteine (L-Cys). The basic reaction involved acetylthiocholine (ATCh) hydrolysis catalysed by acetylcholinesterase resulting in the formation of thiocholine (Tch), which further reacts with DTNB to form TNB. The so formed TNB is capable of causing BPQD fluorescence quenching. The developed fluorescent biosensor established significant efficiency and appreciable sensing limit. The sensor was established according to inner filter effect that utilizes absorbent to quench the fluorescence of fluorophores [63].

5.8 Calorimetric Biosensors

BP possesses immense potential for calorimetry based biomarker detection. Though the low stability of BP obstruct its optical and electrical applications but the same degradation tendency make positive contribution to colorimetric detection sensitivity. Easy degradation tendency confirms the readiness to get reduced and appreciably high electron- donor tendency. These two properties are very much essential for developing a calorimetric biosensor and therefore BP turn out to be an excellent option. The narrow band gap and redox potential of BP facilitate its electron donor tendency. In a recent study, few layer BP- gold nanoparticles hybrid was demonstrated to be highly sensitive calorimetric biosensor for identifying cancer biomarkers. The developed hybrid acted as efficient catalyst reduction of 4-nitrophenol. 4-nitrophenol which is yellow turns colourless when reduced to 4-aminophenol, cancer biomarker. Oxidation of BP facilitate electron transfer that enhances the reduction of 4-nitrophenol to 4-aminophenol. Conjugation of carcinoembryonic antibody onto the BP-Au nanoparticle hybrid surface, mask the catalytic activity of hybrid. In the presence of carcinoembryonic antigen, the catalytic activity is regained since the conjugated carcinoembryonic antibody escapes from the surface and form antibody-antigen complex. This calorimetric reaction can be effectively developed for cancer diagnosis.

The analysis of blood serum from breast cancer and colon cancer patients revealed that the so developed calorimetric sensor showed a remarkable sensitivity and long detection range. The results of clinical blood analysis by the developed calorimetric sensor suggest that BP-Au hybrid can be used for cancer detection notably for colon and breast cancer. The high electron transfer tendency and electron-donor facility gives enhanced detection limit to sense cancer biomarkers [64].

5.9 Chemiluminescent Biosensors

Electrodes modified with BPQDs of size 8.2 nm were found to be exceptionally good electrogenerated chemiluminescence (ECL) platform. ECL is a reliable technique for electrochemical biosensor applications and various systems are examined to attain a good ECL platform. $Ru(bpy)_3^{2+}$ is one such ECL platform explored which usually require a co-reactor. Most of the reported co-rectors suffer limitation of low light emission. Toxicity and poor water solubility are other issues associated with most of the co-reactors for ECL platform. But it was found that BPQDs exhibited strong ECL emission. BPQDs put forward admirable catalytic activity for $Ru(bpy)_3^{2+}$ oxidation to $Ru(bpy)_3^{3+}$. The inhibitory potential of dopamine on ECL emission facilitates its detection using BPQDs based chemiluminescent biosensors. Dopamine when oxidized react with $Ru(bpy)_3^{2+}$. This imparts a competition between $Ru(bpy)_3^{2+}$ oxidation reaction and oxidized dopamine product- $Ru(bpy)_3^{2+}$ leading to decrease in ECL signal [65].

5.10 Imaging Applications

Nanoparticles are widely explored for facilitating imaging purposes. Among various bioimaging techniques, photoacoustic (PA) imaging is a relatively new one with finer sensitivity and resolution. PA put forward good quality contrast images that could be used for imaged assisted disease diagnosis and treatment especially in case of tumor. But early tumor sites produce weak signals making early cancer detection tricky. This hurdle can be overcome using nanoparticles capable of adsorbing infrared radiations and act as contrast agents. The exceptional optical properties make BP good for imaging. Miao et al. recent developed BP based camera for near-IR imaging. The fabricated BP based photodetector based on compressive sensing algorithm produced laser spot images. The results were promising evidence for use of BP based nanodevices for imaging applications [66].

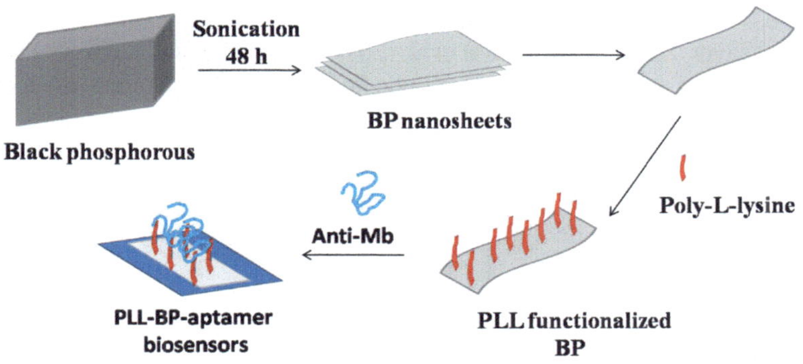

Fig. 5 BP functionalized with aptamer for electrochemical sensing of myoglobin

5.11 Cardiovascular Applications

Cardiovascular diseases account for major proportion of the total death from non-communicable diseases. It is therefore highly essential to have facility to monitor cardiac functioning and diagnose any variation in signals during cardiac events. Cardiac biomarkers plays important role in identifying dysregulation in cardiac functions. Myoglobin (Mb) acts as an excellent cardiac marker especially in diagnosis of acute myocardial infarction (AMI). Mb is found to be over expressed at times of heart attack. Therefore, identifying higher concentration of Mb in blood help in rapid diagnosis and treatment of heart attack [67]. Biochemical sensor based point-of-care (POC) is a reliable option as cardiac biomarker [68]. Kumar V et al. developed poly-L-Lysine functionalized BP conjugated with aptamer as nanostructured electrodes for electrochemically sensing of Mb. The developed aptasensor platform exhibited remarkable sensitivity and record sensing capability toward Mb with high detectable value [69]. The developed BP based nanoplatform provides possibility for translating bench side research to bedside diagnosis of cardiovascular diseases in humans (Fig. 5).

5.12 Molecular Medicine

Personalized medicine can potentially revolutionize the current health care system. For accomplishing personalized medicine based patient care, it is mandatory to have highly efficient as well as innovative DNA sequencing assisted devices [70]. Nanopore/nanogap aided DNA sequencing is an emerging DNA sequencing device capable of highly precise DNA sequencing at a cheaper and faster rate [71]. In the background, solid-state nanopore-aided techniques is reported to attain each nucleotide sequencing with high resolution, superior stability and sequencing speed as compared to biological pores [72]. The DNA sequencing devices depending on nanopore were developed using graphene because of the possibility to have single atom thick layer. Recently, Kumawat et al. (2019) reported that it is possible to identify all the human nucleotides by using single layered BP (phosphorene) as electrode. The findings suggest that phosphorene based electrode can be explored for developing nanodevices for sequencing of DNA and there by utilizing the technique to detect important diseases [73].

6 Safety of BP Nanodevices

Toxicity studies regarding materials originating from elemental phosphorus has previously been concentrating on white and red phosphorus which are the two widely known allotropes. The former is reported to be virtually toxic as cyanide; eliciting unpleasant variances like bone necrosis for the duration of long episodes. According to reports, inhalation of white phosphorus for a period of 5–15 min may contribute to severe health consequences [74]. Meanwhile, red phosphorus is comparatively less toxic and is accepted to be applied as an effective photocatalyst for environmental applications [75]. A generalized proclamation on safety aspects of BP could not be drawn till now. Nevertheless, non-toxic behaviour of BP has been reported for which a suitable supporting data is not available so far. Latiff et al. in 2015 have conducted an in vitro toxicity assessment for BP using lung carcinoma cell line A549. Cytotoxicity level was primarily evaluated using WST-8 and MTT assays. Cell viability was evaluated by exposing cells with different concentrations of BP (3.125–400 μg/ml) for 24 h. The viability pattern was similar for both the assays. Up to the concentration of 25 μg/ml, the percentage of viability was observed to be decreasing. The initial trend made definite that BP is toxic with lowest viability value reaching 48% and 34% for both the assays respectively. Strikingly beyond the level of higher concentration, the cell viability was higher which demands the need to study particle interference with assay reagents. Further study with ascorbic acid revealed that BP interferes greatly with both WST-8 and MTT reagents. Hence these assays are recommended least for the cytotoxicity assessments for BP [76].

Mo et al. in 2018 have exploited protein corona formation on BP nanomaterials for studying macrophage mediated immune level activities. Influence of particle size on formation of protein corona formation was also analyzed by allowing reaction between different sized BP nanomaterials and plasma proteins. BPQDs and BPNS were compared in terms of protein corona and their composition. Protein bands formed on the surface of the particles were analyzed after incubating with plasma proteins of different concentrations via gel electrophoresis. According to densitometry analysis, protein bands were appeared in accordance with the size of the BP NPs. According to mass spectroscopic analysis, 69.9% of total proteins were found on the surface of BPNS whereas BPQDs enclosed 75.8%.

Moreover it was noticeable that among those adsorbed onto BPNS, 52 proteins were found to from plasma and BPQDs consisted of 96 plasma proteins. Cytotoxic potential of the particles was then analyzed by means of MTT assay using dTHP-1cells. Both nanomaterials showed cell viability of >90% for 6 h of incubation. Cellular uptake of both the materials was evaluated using inductively coupled plasma mass spectroscopy (ICPMS) and it was found that BPQDs exhibited more content of phosphorus than that of BPNS. Triggering of ROS generation was analyzed using Dichlorofluorescein (DCF) fluorescence assay and significant release of ROS was noticed in a time depended manner. Overall study accentuates BP nanomaterials exhibit immune induction in macrophages via protein corona formation on their surface [77].

On the contrary, reports highlighting non-toxicity of BPNS in vitro using different cell lines and blood cells are also available. Chen et al. in 2017 have analyzed the toxicity of BPNS in cells like 4T1, HeLa, L929 and A549 at a concentration of 200 μg/ml. In addition, haemolytic potential of BPNS was evaluated in RBC after incubation for 8 h at 37 °C. Almost no signs of cytotoxicity were noticed even at the concentration of 200 μg/ml. In vivo toxicity studies using healthy Sprague-Dawley rats were also conducted. Biochemistry parameters including RBCs, WBCs, platelets and Haemoglobin levels were assessed. None of the tested parameters showed any alterations. On 7th day of post exposure, clinical biochemistry parameters were evaluated (alanine aminotransferase, aspartate aminotransferase, urea, creatinine, total proteins etc.); as a mean to analyze tissue functionalities. A significant change in tissue functions was not evident during observation. Animals neither developed any alterations with body or organ weight nor histopathological lesions. In this way, the particular study revealed the superior biosafety potentials of BPNS.

Even though cytotoxicity studies for BP nanomaterials are relatively insufficient, available in vitro and in vivo studies show that BP is a safe material to be utilized in various medical and non-medical applications even in a long term basis. However, several modification strategies could further improve the quality to a significant level which may definitely increase the commercial requirements [78].

7 Challenges and Refinement Strategies

In spite of the attractive properties of black phosphorous the difficulty in producing large area BP layers and rapid surface degradation of BP limits its transformation to medical devices. Mechanical exfoliation is the most relied exfoliation technique to produce BP layers from bulk. But though mechanically exfoliated BP layers accomplish the quality requirements, the difficulty in scaling up the production further hinders the large scale production of bigger BP layers. Liquid exfoliation technique is another dependable option to BP monolayer fabrication. Liquid exfoliation helps scaling up of BP production but the so produced BP flakes lack crystallinity that confines its applications in certain electronic and optical applications. More advanced exfoliation techniques are now been developed so as to produce higher amount of superior quality larger area BP layers required to develop proposed BP based nanodevices. Chemical vapour deposition (CVD) technique is one such promising method capable of producing high quality BP layers with high surface area and controlled thickness [79]. But CVD is less explored and comparatively limited information is available. Other exfoliation techniques like plasma etching and laser irradiation still remain unsuitable for large scale production of BP layers due to cost ineffectiveness [80]. More reliable fabrication techniques capable of producing large surface BP layers with tunable thickness and finer quality in a cost effective way is still a prime target in BP research.

The surface degradation is resulted from the ROS due to photo-oxidation at the surface [81]. Oxygen is readily adsorbed on to the BP surface quickly resulting in atomic level defects. Atomic defects further induce oxidation by raising the lattice spacing. The atomic level neutral defects get more pronounced as the number of layers decreases. In monolayers since most of the atoms reside at the surface itself number of defects/atom is higher compared to bulk. In monolayers the defects spreads from edges to the middle portions resulting in more predominant degradation [82]. The adsorption of water molecules is decreases with increasing layer thickness. Therefore monolayers show more advanced degradation due to enhanced wettability.

Also with the drop in layer thickness, the band width changes to oxygen accepting range resulting in enhanced oxidation. Oxygen in the presence of light attains targeted activation energy of 0.69 eV and penetrates through the BP lattice causing crystal degradation resulting in formation of phosphorus trioxide or phosphorus pentoxide [83]. Island et al. evaluated the contribution of environmental factors in effecting the efficiency of FETs. According to the study, the adsorption of oxygen and nitrogen moieties resulted in the shift in threshold voltage causing changes in BP FET transfer characteristics in small time scale. A change in threshold voltage from left to right resulting in decrease in conductance is observed on BP exposure to air. Long timescale changes in FET characteristic include p-type doping due to water adsorption. BP based FET measurements over long term air exposure showed degradation and breaking down of channel as a result of layer-by layer etching [84].

Addressing the above mentioned two challenges is mandatory call so as to translate the bench top research on BP to effective commerciable nanodevices. Inorder to stabilize BP from environmental factors two protection strategies namely physical means and chemical processes are developed. In both ways the idea is to make BP inert by cut-shorting the exposure towards various environmental factors. The chemical methods of protection involve surface treatments so as to avoid exposure to environmental factors and ambient conditions that contribute towards degradation. The physical route engrosses passivation of BP by incorporating protective layers [85]. Although much attention has been devoted on strategies to tackle BP degradation, finding a reliable and effective protection method still remains a challenge. One of the generally used physical ways is to encapsulate BP with dielectric materials. Al_2O_3 is the most explored dielectric material as BP capping agent. Even though capping with Al_2O_3 can produce BP stable up to a time span of 17 months, the charge entrapping at high temperature associated with Al_2O_3 depositions effect the performance for FET applications [86].

The failure to hold the FET characteristics of BP and degradation in performance associated with encapsulation calls for more reliable techniques to overcome the degradation. Another limitation associated with passivation is the requirement of techniques like atomic layer deposition (ALD) which fails to achieve uniform deposition. Passivation of single layer BP/phosphosene by ALD techniques is extremely difficult because of the quick oxidation. More advanced techniques like plasma enhance atomic layer deposition is now explored to achieve uniform deposition of dielectric material on to BP surface [87].

Interestingly, various 2D material such as graphene and h-BN are used for encapsulating BP. Encapsulating with h-BN require more difficult conditions and result in lower yield making it less reliable option [88]. The chemical approaches for minimizing BP degradation include surface functionalisation/doping. Chemical functionalisation take advantage of the bonding between the BP and the functionalizing agent used so as to suppress the degradation. Diazonium salts are widely used for functionalisation of BP since covalent bond formation between the carbon atom of diazonium salt and P atom of BP helps in suppression of BP degradation [89]. Similarly, conjugation of $TiCl_4$ ligand with BP results in formation of coordination bond between Ti and P utilizing the lone pair of p atom in BP thereby reducing chances of degradation [90]. All the above mentioned techniques utilized the principle of minimizing the exposure of BP towards air as well as moisture. Using imidazolium-based ionic liquids on BP surface overcomes degradation. The study employed [BMIM][BF4] ionic liquids helped imidazolium-based ionic liquids in quenching reactive oxygen species (ROS) thereby evade photo-oxidative damage [91]. Another reliable chemical approach for attaining stability is by doping different elements onto BP. Doping with Te is also reported to boost the stability of BP. Development of more convenient techniques to augment stability is of high priority. Researchers have been involved in addressing the above mentioned challenges and are focused on finding solution.

The use of roll coating method for fabrication of BP based electrodes hamper scaling up production thereby restricting BP to laboratory research. Conversion of BP research from laboratory to marketable BP based nanodevices need to be accomplished at a cost-effective means by overcoming all the associated hurdles. Even though many difficulties persist before practical implication, BP based nanodevices are worth consideration for various electronic as well as biomedical applications.

8 Conclusion

The surge for 2D materials has increased to a considerable level nowadays markedly due to the residing inimitable characteristics. In addition, higher band gap, strong structural and functional anisotropy, higher conductivity etc. have turned it to be fascinating material among researchers. Being a growing idol in nanotechnology field, BP based nanomaterials and devices play inevitable role in achieving tremendous growth in this scenario. Studies got flourished with the peculiar role of BP based materials over other 2D materials which could definitely form a bridge between other interdisciplinary areas of science and technology. The chapter emphasized more on current status of BP materials in global and Indian scenarios, different types of BP, applications of BP in non-medical and medical fields, their safety and challenges observed so far. Non-medical applications using BP nanomaterials discussed here are: as gas sensors, water splitting photocatalysts, as energy storage device, memory devices etc. BP has been found to possess roles in energy storage and conversion. However, high reactivity towards oxygen characteristically because of the presence of lone pair of electrons hinders their application potentials to a larger extend. Such interactions results in degradation of the material into phosphoric acid and PO_4^{3-} ions which makes the material unstable under ambient conditions. Despite the admirable properties, the translational research applications of the material are debatable; mostly because of the poor stability in air and aqueous environment. These further necessitate functionalization strategies to be employed or fabrication of hetero structures with other 2D materials.

One of the intriguing application potentials of BP includes higher photothermal ability which could be exploited for localized hyperthermia in cancer theranostics. Characteristic generation of $^1O_2$ by BP photocatalyst can be utilized for curing dreadful diseases. BPNS assisted combined strategies like PTT and PDT have found remarkable success in ablation of cancers to date. Because of the inherent peculiarities of different aspects, BP has the ability to cross through biological barriers including BBB. In this way it can reach the brain loci and capture redox active metal ions; thereby performing neuroprotective roles. BP based 3D scaffolds hold the potential of osteoinduction which have found abundant potentials in bone regeneration. Higher safety level and degradation potential of BP in realistic environment have been shown by several biocompatibility and toxicity studies. However, still there occurs a wide gap for generalization regarding toxicity levels. Moreover conventionally used assays can longer be applied for BP because of the undesirable interference with assay reagents including MTT. This further demands suitable alternatives. Still according to available literatures, BP exhibits relatively lower toxicity in various cell lines tested. In vivo studies are also found to be in conjunction with these non-toxicity reports. Therefore as a whole it is appreciable to state that BP based nanomaterials requires more deepened research in various aspects of efficacy, toxicity, modifications as well as therapeutic potentials. The hidden possibilities of applications potentials are needed to get unveiled for the nurturing of medical as well as non-medical fields.

Acknowledgements The authors wish to express their sincere thanks to Director and Head, Biomedical Technology Wing, Sree Chitra Tirunal Institute for Medical Sciences and Technology, Thiruvananthapuram for their encouragement and support for conducting this study. Ashtami thanks SCTIMST, Trivandrum, Athira thanks CSIR, New Delhi and Reshma thanks DST (Inspire fellowship), New Delhi for financial support of Junior Research Fellowships.

Conflict of Interest
The authors declare that they have no conflict of interest.

References

1. Ramsden, J.J.: Applied Nanotechnology: The Conversion of Research Results to Products, A volume in Micro and Nano Technologies Book, 2nd edn. (2014)
2. Bridgman, P.W., et al.: Two new modification of phosphorous. J. Am. Chem. Soc. **36**(7), 1344–1363 (1914)
3. Rudenko, A.N., Katsnelson, M.I.: Quasiparticle band structure and tight-binding model for single- and bilayer black phosphorus. Phys. Rev. B **89**(20), 201408 (2014)
4. Lee, T., Kim, S., et al.: Black phosphorus: critical review and potential for water splitting photocatalyst. Nanomaterials **6**(11), 194 (2016)
5. Kikegawa, K., Iwasaki, H.: An X-ray diffraction study of lattice compression and phase transition of crystalline phosphorus. Acta Crystallogr. Sect. B **39**, 158–164 (1983)
6. Rodin, A., Carvalho, A., Neto, A.C., et al.: Strain-induced gap modification in black phosphorus. Phys. Rev. Lett. **112**, 176801 (2014)
7. Rodin, A.S., Carvalho, A., et al.: Excitons in anisotropic twodimensional semiconducting crystals. Phys. Rev. B **90**(7), 075429 (2014)
8. Buscema, M., Groenendijk, D.J., et al.: Fast and broadband photoresponse of few-layer black phosphorus field-effect transistors. Nano Lett. **14**(6), 3347–3352 (2014)
9. Castellanos-Gomez, A., Vicarelli, L., et al.: Isolation and characterization of few-layer black phosphorus. 2D Mater. **1**(2), 025001 (2014)
10. Koenig, S.P., Rostislav, A., et al.: Electric field effect in ultrathin black phosphorus. Appl. Phys. Lett. **104**(10), 103106 (2014)
11. Li, L., Yijun, Yu., et al.: Black phosphorus field-effect transistors. Nat. Nanotechnol. **9**(5), 372–377 (2014)
12. Liu, H., Neal, A.T., Zhu, Z., et al.: Phosphorene: an unexplored 2D semiconductor with a high hole mobility. ACS Nano **8**, 4033–4041 (2014)
13. Xia, F., Wang, H., et al.: Rediscovering black phosphorus as an anisotropic layered material for optoelectronics and electronics. Nat. Commun. **5**, 4458 (2014)
14. Buscema, M., Dirk, J., et al.: Photovoltaic effect in few-layer black phosphorus PN junctions defined by local electrostatic gating. Nat. Commun. **5**, 4651 (2014)
15. Deng, Y., Luo, Z., et al.: Black phosphorus-monolayer MoS_2 van der Waals heterojunction P-N diode. ACS Nano **8**(8), 8292–8299 (2014)
16. Wang, L., Sofer, Z., Pumera, M.: Voltammetry of layered black phosphorus: electrochemistry of multilayer phosphorene. Chem. Electro. Chem. **2**(3), 324–327 (2015)
17. Castellanos-Gomez, A.: Black phosphorus: narrow gap, wide applications. J. Phys. Chem. Lett. **6**(21), 4280–4291 (2015)
18. Xiong, X., Li, X.: High performance black phosphorus electronic and photonic devices with HfLaO dielectric. IEEE Electron Dev. Lett. **39**(1), 127–130 (2018)
19. Zhu, W.: Black phosphorus thin-film transistors: from strain tunability to high frequency applications. Doctoral dissertation (2017)
20. Chen, H., Huang, P., et al.: Anisotropic mechanical properties of black phosphorus nanoribbons. J. Phys. Chem. C **120**(51), 29491–29497 (2016)
21. Zhu, W., Park, S., et al.: Black phosphorus flexible thin film transistors at gighertz frequencies. Nano Lett. **16**(4), 2301–2306 (2016)
22. Zhu, W., Maruthi, N., et al.: Flexible black phosphorus ambipolar transistors, circuits and AM demodulator. Nano Lett. **15**(3), 1883–1890 (2015)
23. Tran, R.S., Liang, Y., et al.: Layer-controlled band gap and anisotropic excitons in few-layer black phosphorus. Phys. Rev. B **89**(23), 235319 (2014)
24. Cai, Y., Zhang, G., et al.: Layer-dependent band alignment and work function of few-layer phosphorene. Sci. Rep. **4**, 6677 (2014)
25. Das, S., Demarteau, M., et al.: Ambipolar phosphorene field effect transistor. ACS Nano **8**(11), 11730–11738 (2014)
26. Chang, Hsiao-Yu., Yang, S., et al.: High-performance, highly bendable MoS_2 transistors with high-k dielectrics for flexible low-power systems. ACS Nano **7**(6), 5446–5452 (2013)

27. Abbas, A.N., Liu, B., Chen, L., et al.: Black phosphorus gas sensors. ACS Nano **9**(5), 5618–5624 (2015)
28. Cui, S., Pu, H., Wells, S.A., et al.: Ultrahigh sensitivity and layer-dependent sensing performance of phosphorene-based gas sensors. Nat. Commun. **6**, 8632–8641 (2015)
29. Kou, L., Frauenheim, T., et al.: Phosphorene as a superior gas sensor: selective adsorption and distinctI–V response. J. Phys. Chem. Lett. **5**, 2675–2681 (2014)
30. Yan, S., Wang, B., et al.: Supercritical carbon dioxide-assisted rapid synthesis of few-layer black phosphorus for hydrogen peroxide sensing. Biosens. Bioelectron. **80**, 34–38 (2016)
31. Liu, B., Kopf, M., Abbas, A.N., et al.: Black arsenic-phosphorus: layered anisotropic infrared semiconductors with highly tunable compositions and properties. Adv. Mater. **27**, 4423–4429 (2015)
32. Dai, J., Zeng, X.C.: Bilayer phosphorene: effect of stacking order on bandgap and its potential applications in thin-film solar cells. J. Phys. Chem. Lett. **5**, 1289–1293 (2014)
33. Sa, B., Li, Y.-L., et al.: Strain engineering for phosphorene: the potential application as a photocatalyst. J. Phys. Chem. C **118**(46), 26560–26568 (2014)
34. Chen, W., Li, K., Wang, Y., et al.: Black phosphorus quantum dots for hole extraction of typical planar hybrid perovskite solar cells. J. Phys. Chem. Lett. **8**(3), 591–598 (2017)
35. Ramireddy, T., Xing, T., Rahman, M.M., et al.: Phosphorus–carbon nanocomposite anodes for lithium-ion and sodium-ion batteries. J. Mater. Chem. A **3**(10), 5572–5584 (2015)
36. (a) Zhang, J., Liu, H.J., Cheng, L.: High thermoelectric performance can be achieved in black phosphorus. J. Mater. Chem. C **4**(5), 991–998 (2016); (b) Zhou, Y., Zhang, M., Guo, Z., et al.: Recent advances in black phosphorus-based photonics, electronics, sensors and energy devices. Mater. Horiz. **4**(6), 997–1019 (2017)
37. An, C.J., Kang, Y.H., Lee, C., et al.: Preparation of highly stable black phosphorus by gold decoration for high-performance thermoelectric generators. Adv. Funct. Mater. **28**(28), 1800532–1800539 (2018)
38. Yan, Z.-Q., Zhang, W.: The development of graphene-based devices for cell biology research. Front. Mater. Sci. **8**(2), 107–122 (2014)
39. Wood, J.D., Wells, S.A., et al.: Effective passivation of exfoliated black phosphorus transistors against ambient degradation. Nano Lett. **14**(12), 6964–6970 (2014)
40. Li, P., Zhang, D., et al.: Air-stable black phosphorus devices for ion sensing. ACS Appl. Mater. Interfaces **7**(44), 24396–24402 (2015)
41. Zhang, W., Huynh, T., et al.: Revealing the importance of surface morphology of nanomaterials to biological responses: adsorption of the villin headpiece onto graphene and phosphorene. Carbon **94**, 895–902 (2015)
42. Xuzhu, W., Shao, J., Raouf, M.A.E., et al.: Near-infrared light-triggered drug delivery system based on black phosphorus for in vivo bone regeneration. Biomaterials **179**, 164–174 (2018)
43. Song, B., Li, K., et al.: Tuning mixed nickel iron phosphosulfide nanosheet electrocatalysts for enhanced hydrogen and oxygen evolution. ACS Catal. **7**(12), 8549–8557 (2017)
44. Zhou, Z.L., Ying, M., et al.: Black phosphorus nanosheets for rapid microRNA detection. Nanoscale **10**(11), 5060–5064 (2018)
45. Lei, W., Liu, G., et al.: Black phosphorus nanostructures: recent advances in hybridization, doping and functionalization. Chem. Soc. Rev. **46**(12), 3492–3509 (2017)
46. Shao, J., Xie, H., et al.: Biodegradable black phosphorus-based nanospheres for in vivo photothermal cancer therapy. Nat. Commun. **7**, 12967 (2016)
47. Qiu, M., Wang, D., et al.: Novel concept of the smart NIR-light–controlled drug release of black phosphorus nanostructure for cancer therapy. Proc. Natl. Acad. Sci. **115**(3), 501–506 (2018)
48. Sun, C., Wen, L., et al.: One-pot solventless preparation of PEGylated black phosphorus nanoparticles for photoacoustic imaging and photothermal therapy of cancer. Biomaterials **91**, 81–89 (2016)
49. Yang, G., Liu, Z., et al.: Facile synthesis of black phosphorus–Au nanocomposites for enhanced photothermal cancer therapy and surface-enhanced Raman scattering analysis. Biomater. Sci. **5**(10), 2048–2055 (2017)

50. Subramani, K., Mehta, M.: Nanodiagnostics in microbiology and dentistry. Emerg. Nanotechnol. Dent. **2**, 391–419 (2018)
51. Rogers, G.F., Greene, A.K.: Autogenous bone graft: basic science and clinical implications. J. Craniofac. Surg. **23**(1), 323–327 (2012)
52. Yang, B., Yin, J.: 2D-black-phosphorus-reinforced 3D-printed scaffolds: a stepwise countermeasure for osteosarcoma. Adv. Mater. **30**(10), 1705611 (2018)
53. Lu, Y., Alex, A., et al.: Bioresponsive materials. Nat. Rev. Mater. **2**(1), 16075 (2017)
54. Chen, W., Ouyang, J., et al.: Black phosphorus nanosheets as a neuroprotective nanomedicine for neurodegenerative disorder therapy. Adv. Mater. **30**(3), 1703458 (2018)
55. Schedin, F., Geim, A., Morozov, S., et al.: Detection of individual gas molecules adsorbed on graphene. Nat. Mater. **6**, 652–655 (2007)
56. Wang, Z., Feng, P.X.L.: Design of black phosphorus 2D nanomechanical resonators by exploiting the intrinsic mechanical anisotropy. 2D Mater. **2**(2), 021001 (2015)
57. Mayorga-Martinez, C.C., Mohamad Latiff, N., Eng, A.Y.S., et al.: Black phosphorus nanoparticle labels for immunoassays via hydrogen evolution reaction mediation. Anal. Chem. **88**(20), 10074–10079 (2016)
58. Chen, Y., Ren, R., Pu, H.: Field-effect transistor biosensors with two-dimensional black phosphorus nanosheets. Biosens. Bioelectron. **89**, 505–510 (2017)
59. Li, P., Zhang, D., Jiang, C., et al.: Ultra-sensitive suspended atomically thin-layered black phosphorus mercury sensors. Biosens. Bioelectron. **98**, 68–75 (2017)
60. Cho, S.Y., Lee, Y., Koh, H.J., et al.: Superior chemical sensing performance of black phosphorus: comparison with MoS_2 and graphene. Adv. Mater. **28**(32), 7020–7028 (2016)
61. Niu, X., Li, Y., Shu, H., et al.: Anomalous size dependence of optical properties in black phosphorus quantum dots. J. Phys. Chem. Lett. **7**, 370–375 (2016)
62. Yew, Y.T., Sofer, Z., Mayorga-Martinez, C.C., et al.: Black phosphorus nanoparticles as a novel fluorescent sensing platform for nucleic acid detection. Mater. Chem. Front. **1**, 1130–1136 (2017)
63. Gu, W., Yan, Y., Pei, X., et al.: Fluorescent black phosphorus quantum dots as label-free sensing probes for evaluation of acetylcholinesterase activity. Sens. Actuators B Chem. **250**, 601–607 (2017)
64. Peng, J., Lai, Y., Chen, Y., et al.: Sensitive detection of carcinoembryonic antigen using stability-limited few-layer black phosphorus as an electron donor and a reservoir. Small **13**(15), 1603589–1603600 (2017)
65. Zhang, L., Tian, K., Dong, Y., et al.: Electrogenerated chemiluminescence of Ru $(bpy)_3^{2+}$ at a black phosphorus quantum dot modified electrode and its sensing application. Analyst **143**(1), 304–310 (2018)
66. Miao, J., Song, B., Xu, Z., et al.: Single pixel black phosphorus photodetector for near-infrared imaging. Small **14**(2), 1702082–1702089 (2018)
67. Moreira, F.T.C., Dutra, R.A.F., Noronha, J.P.C., et al.: Electrochemical biosensor based on biomimetic material for myoglobin detection. Electrochim. Acta **107**, 481–487 (2013)
68. Yang, Z., Zhou, D.M.: Cardiac markers and their point-of-care testing for diagnosis of acute myocardial infarction. Clin. Biochem. **39**(8), 771–780 (2006)
69. Kumar, V., Brent, J.R., Shorie, M., et al.: Nanostructured aptamer-functionalized black phosphorus sensing platform for label-free detection of myoglobin, a cardiovascular disease biomarker. ACS Appl. Mater. Interfaces **8**(35), 22860–22868 (2016)
70. Heerema, S.J., Dekker, C.: Graphene nanodevices for DNA sequencing. Nat. Nanotechnol. **11**, 127–136 (2016)
71. Branton, D., Deamer, D.W., Marziali, A., et al.: The potential and challenges of nanopore sequencing. Nat. Biotechnol. **26**, 1146–1153 (2008)
72. Wanunu, M.: Nanopores: a journey towards DNA sequencing. Phys. Life Rev. **9**, 125–158 (2012)
73. Kumawat, R.L., Garg, P., Kumar, S., et al.: Electronic transport through DNA nucleotides in atomically thin phosphorene electrodes for rapid DNA sequencing. ACS Appl. Mater. Interfaces **11**(1), 219–225 (2018)

74. Nitschke, J.R.: The two faces of phosphorus. Nat. Chem. **3**(1), 90 (2010)
75. Xia, D., Shen, Z., et al.: Red phosphorus: an earth-abundant elemental photocatalyst for "green" bacterial inactivation under visible light. Environ. Sci. Technol. **49**(10), 6264–6273 (2015)
76. Latiff, N.M., Teo, W.Z., et al.: The cytotoxicity of layered black phosphorus. Chem. Eur. J. **21**(40), 13991–13995 (2015)
77. Mo, J., Xie, Q., et al.: Revealing the immune perturbation of black phosphorus nanomaterials to macrophages by understanding the protein corona. Nat. Commun. **9**(1), 2480 (2018)
78. Chen, W., Ouyang, J., et al.: Black phosphorus nanosheet-based drug delivery system for synergistic photodynamic/photothermal/chemotherapy of cancer. Adv. Mater. **29**(5), 1603864 (2017)
79. Smith, J.B., Hagaman, D., Ji, H.F.: Growth of 2D black phosphorus film from chemical vapor deposition. Nanotechnology **27**(21), 215602–215610 (2016)
80. Wu, S., Hui, K.S., Hui, K.N.: 2D black phosphorus: from preparation to applications for electrochemical energy storage. Adv. Sci. **5**(5), 1700491–1700519 (2018)
81. Zhou, Q., Chen, Q., Tong, Y.: Light induced ambient degradation of few layer black phosphorus: mechanism and protection. Angew. Chem. Int. Ed. **55**(38), 11437–11441 (2016)
82. Alsaffar, F., Alodan, S., Alrasheed, A., et al.: Raman sensitive degradation and etching dynamics of exfoliated black phosphorus. Sci. Rep. **7**, 44540–44549 (2017)
83. Ziletti, A., Carvalho, A., Campbell, D.K., et al.: Oxygen defects in phosphorene. Phys. Rev. Lett. **114**(4), 046801–046806 (2015)
84. Island, J.O., Steele, G.A., van der Zant, H.S., et al.: Environmental instability of few-layer black phosphorus. 2D Mater. **2**(1), 011002–011009 (2015)
85. Kuriakose, S., Ahmed, T., Balendhran, S., et al.: Black phosphorus: ambient degradation and strategies for protection. 2D Mater. (3), 032001–032025 (2018)
86. Illarionov, Y.Y., Waltl, M., Rzepa, G., et al.: Highly-stable black phosphorus field-effect transistors with low density of oxide traps. NPJ 2D Mater. Appl. **1**(1), 23–30 (2017)
87. Wu, B.B., Zheng, H.M., Ding, Y.Q., et al.: Direct growth of Al_2O_3 on black phosphorus by plasma-enhanced atomic layer deposition. Nanoscale Res. Lett. **12**(1), 282–288 (2017)
88. Cai, Y., Zhang, G., Zhang, Y.W.: Electronic properties of phosphorene/graphene and phosphorene/hexagonal boron nitride heterostructures. J. Phys. Chem. C **119**(24), 13929–13936 (2015)
89. Ryder, C.R., Wood, J.D., Wells, S.A., et al.: Covalent functionalization and passivation of exfoliated black phosphorus via aryl diazonium chemistry. Nat. Chem. **8**(6), 597–602 (2016)
90. Zhao, Y., Wang, H., Huang, H., et al.: Surface coordination of black phosphorus for robust air and water stability. Angew. Chem. **55**(16), 5003–5007 (2016)
91. Walia, S., Balendhran, S., Ahmed, T., et al.: Ambient protection of few-layer black phosphorus via sequestration of reactive oxygen species. Adv. Mater. **29**(27), 1700152–1700160 (2017)

Chemistry of Black Phosphorus

Mohammad Ghashghaee, Mehdi Ghambarian and Zahra Azizi

Abstract Black phosphorus (BP), a rediscovered one-element two-dimensional (2D) nanomaterial, has been intensively explored in the past few years. Of particular interest is its single-layer structure, called phosphorene. Understanding the chemistry of BP can substantially help in the development of BP-based practical devices. This chapter provides an overview of various aspects of BP chemistry.

Abbreviations

2D	Two-dimensional
AFM	Atomic force microscopy
ARPES	Angle-resolved photoemission spectroscopy
BE	Binding energy
BCS	Bardeen–Cooper–Schrieffer
BP	Black phosphorus
h-BN	Hexagonal boron nitride
PDI	Perylene bisimide
PL	Photoluminescence
RP	Red phosphorus
RPA	Random phase approximation
TCNQ	Tetracyano-*p*-quinodimethane
TMD	Transition metal dichalcogenide

M. Ghashghaee
Faculty of Petrochemicals, Iran Polymer and Petrochemical Institute, P.O. Box 14975-112, Tehran, Iran

M. Ghambarian
Gas Conversion Department, Faculty of Petrochemicals, Iran Polymer and Petrochemical Institute, P.O. Box 14975-112, Tehran, Iran

Z. Azizi (✉)
Department of Chemistry, Karaj Branch, Islamic Azad University, P.O. Box 31485-313, Karaj, Iran
e-mail: zahra.azizi@kiau.ac.ir

Inamuddin et al. (eds.), *Black Phosphorus*, Engineering Materials,
https://doi.org/10.1007/978-3-030-29555-4_3

TOF-ND	Time-of-flight neutron powder diffraction
vdW	Van der Waals
VP	Violet Phosphorus
WF	Wave function
WP	White phosphorus
ZT	The highest figure of merit

1 Introduction

Black phosphorus (BP), which is one of the four well-defined allotropes of P with an attractive semiconducting behavior, was coined in the beginning of the 19th century [1]. Similarly to other van der Waals (vdW) substances, BP is composed of 2D nanosheets, but with in-plane bonding [2–5]. A 100 years after the first preparation in 1914, the advent of the exfoliated single-layer BP (called phosphorene) renewed the attention in rapidly expanding research activities on this old material [3, 6–8]. Knowledge of the BP chemistry is critically important to (1) explore the principles of its intrinsic chemical reactivity, (2) enhance its solubility and processability while avoiding its high oxophilicity, (3) provide concepts for extensive fine-tuning and modulation of its physical properties, (4) establish protocols for combined physical/material properties in composite systems with other compound classes, and (5) establish the fundamental basis for numerous practical applications [9]. The prime reactivity concern about BP, which currently poses limitations to the utilization of its chemistry and promising capabilities is the well-known inherent instability [10] of its few-layer nanostructures after their mechanical [6] or solvent [11] exfoliation [5, 9]. The oxidative decomposition of BP occurs in ambient oxygen and moisture [12], which can be intensified under light irradiation through photodegradation [9, 13]. Concerning the advancements, e.g., some recent studies of the stabilization with the perylene bisimide (PDI) molecules or 7,7,8,8-tetracyano-p-quinodimethane (TCNQ) have appeared as the first works in the realm of supramolecular chemistry of BP nanomaterials [9, 14].

2 Crystal Structure

Phosphorus (P) is one of the most copious elements (comprising about ~0.1%) in the earth crust [5, 7, 15]. Phosphorus exists in different forms of cubic white (WP), amorphous red (RP), monoclinic violet (VP), and orthorhombic black (BP) crystalline structures [7], while BP is the most stable allotrope of P with ignition resistance at ambient conditions [5–7, 15–19]. BP can be synthesized through high-pressure heating of WP. BP has a layered crystal structure (Fig. 1) and appears black and flaky like graphite [20]. The calculated interlayer distance in BP is ~5.3 Å [21]. Such

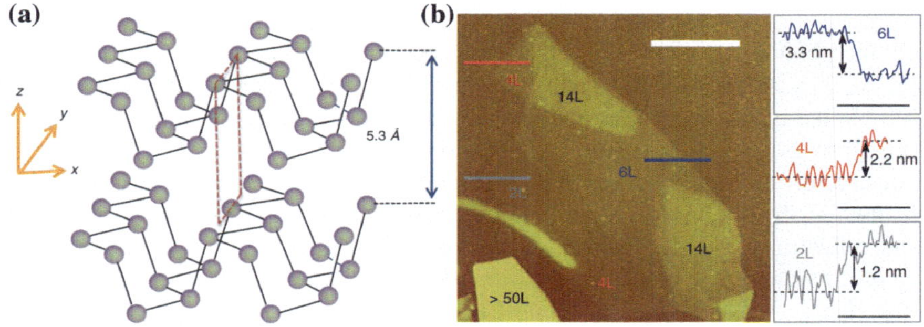

Fig. 1 Layered crystal matrix of BP with **a** the configuration of two buckled layers having an interlayer spacing of about 5.3 Å, and **b** an atomic force microscopic (AFM) image of the flakes on the left with a the thick white scale bar of 2 mm, and the corresponding line scans in the direction of the colored lines drawn in the right panel [18]

a considerable distance implies that these AB-stacked layers are glued together by weak vdW interactions [5, 17, 22]. The theoretically estimated cohesive energy (E_{coh}) of ~20 meV/atom for BP [6, 23] is close to the corresponding values of hexagonal boron-nitride (h-BN) and graphite (~30 meV/atom) [4, 17, 24].

The chemistry of phosphorus is best understood by taking into account the white P_4 molecule, the atoms of which form a tetrahedron composed of six single bonds. Therefore, each P atom in WP is bonded covalently to three P atoms in the vicinity, resulting from the 3p orbitals. However, the single bonds are not able to take a 90° angle as it can be seen for a pure 3p orbital. Instead, one observes arc-like bonds [25] and, therefore, the WP molecule is defined with its instability 6 [3, 17, 26].

On the contrary, the P_4 units in BP are connected to shape continuous layers via break-ing down the individual bonds to form sp^3 hybridization, having bond angles of 96.34° and 103.09° that approach 109.5° for a perfect tetragonal, thus providing higher stability of its crystal network. As every P atom is then bonded covalently to three adjacent P atoms, every p orbital bears a lone pair (Fig. 1). The residual lone pairs protrude at an angle of ~45° relative to the BP plane. Inspired by the graphene designation, the *x* axis is named as the armchair direction while the *y* axis is called the zigzag direction [3, 15, 17, 26].

Unlike some other inorganic 2D structures, e.g., transition metal dichalcogenides (TMDs), the architecture of BP is somewhat simpler [3, 17]. Owing to the sp^3 hybridization, however, BP cannot form flat sheets at the atomic scale as found in case of graphene [2] and it tends to form puckered honeycomb-like structure [27]. The BP layers are vertically stacked by surmountable vdW forces [3–5, 7, 15, 28]. Consequently, each P atom has two adjacent P atoms and every single layer contains two atomic sublayers in which the bond length between the two nearest P atoms (d1 = 2.224 Å) and the corresponding value for the top and bottom atoms (d2 = 2.244 Å) are a little dissimilar. At the same time, d1 and d2 are not so much different due to the covalent bonding of the P 3p orbitals [3, 29, 30].

The layered bulk BP lattice falls into the space group *Cmca* (#64) classification with the point group *mmm* (D_{2h}^{18}) that involves 12 vibrational modes at the G point [26, 30, 31]. A complete list of the 12 vibrational modes can be found elsewhere [30]. Unlike graphene, the individual monolayers of BP have a symmetry reduced from 16 to 8 operators owing to their buckling [30, 32]. Whereas the prime axis in such structures runs along the buckles, the reflection plane σ_h lies inside the *xz* plane. Meanwhile, the Γ point wave functions (WFs) will be either odd or even with respect to the reflection across the σ_h [32]. Hence, the few-layer BP material is classified into two different space groups (*Pbcm* and *Pmna* for an even and odd number of layers, respectively), while sharing the same C_{2h} point group (the symmetries of D_{2h}^{7} and D_{2h}^{11} for an odd and even number of layer stacks, respectively) [26, 30]. As such, the even-numbered layers do not have inversion symmetry. The most frequent stacking in BP is AB, as illustrated in Fig. 1.

The lattice constants of bulk BP have been identified as 4.374, 3.313, and 10.473 Å for *a*, *b*, and *c*, respectively, using the time-of-flight neutron pow-der diffraction (TOF-ND) [33]. Nevertheless, the anisotropy of the in-plane bonding and the inconclusive amounts of the covalent and weak intralayer bonds make it difficult to precisely regenerate the values above through computational modeling [2, 26].

The puckered orthorhombic structure and coupled hinge-like bonding configuration of BP make it different from many 2D materials with hexagonal planar configurations including h-BN, graphene, and TMDs [3, 26]. The buckling of the BP monolayer structure leads to an unusual mechanical behavior, e.g., as a negative Poisson's ratio (−0.027) in the perpendicular direction to the surface under stretching or compression in the *y*-direction [4, 5, 26, 34, 35].

Some other new layered polymorphs/allotropes of phosphorus have also been predicted theoretically (Fig. 2) [17, 36], which include mainly the three β, γ, and δ forms [19] with distinctive electronic properties. For instance, the blue phosphorus has an in-plane hexagonal matrix with an AB layer stacking [36]. Whereas the blue phosphorus has a larger bandgap than BP [37], the γ-polymorph exhibits an

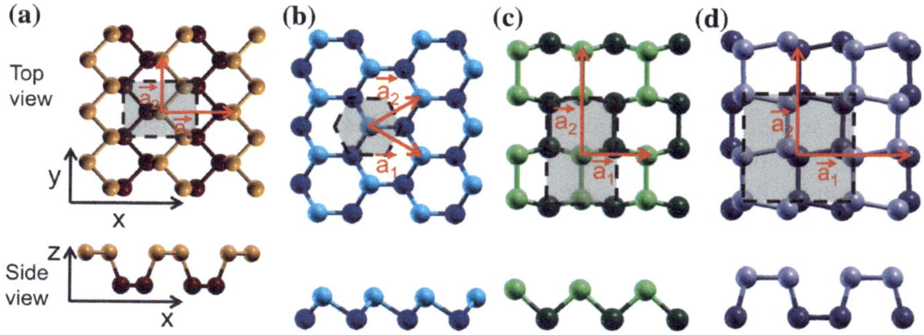

Fig. 2 Top and side views of the monolayer structures of **a** black phosphorus (α-P), **b** blue phosphorus (β-P), **c** γ-phosphorus (γ-P), and **d** δ-phosphorus (δ-P) where the atoms in two adjacent planes are indicated by different color tones [38]

insulator-to-metal transition due to the in-plane strains [38]. The fantastic possibility for connecting different polymorphs of P may provide a unique potential for the assemblage of heterostructures of tunable semiconducting and metallic segments for complex applications [3, 17, 38].

3 Band Structure and Conductivity

The BP electronic band structure shows a direct bandgap (at Z) as also identified by the angle-resolved photoemission spectroscopy (ARPES) data, and ranges from ~0.3 eV (bulk) to ~2.0 eV (single-layer) [3, 17, 25, 26, 35, 39]. Figure 3 shows the electronic band structures of BP. Owing to its band gap, the BP nanoparticles exhibit fluorescence [40]. First-principles calculations have shown that the elimination of interlayer hybridization in few-layer materials is responsible the alteration of their bandgap [3, 6]. The comparative band structures of BP with different numbers of layers have unraveled a redshift of the bandgap and a coincident band splitting as the layer number increased. Meanwhile, the band structure dispersion is kept almost intact. The bandgap appears at the Γ-point (or G-point). However, the upper valence band is approximately flat, and its off-centered maximum (if any) is too shallow to be analyzed at room temperature [2, 17, 26].

The direct bandgap sets the BP monolayer aside from 2D semiconductors such as MoS_2 and WS_2, which exhibit a bandgap transition from direct to indirect upon the build-up of the layers [7, 26, 42]. The same characteristic is shared by less common chalcogenides, e.g. TiS_3, InSe, ReS_2, etc. however; this can be taken as an advantage in optoelectronic applications as it makes it more straightforward to fabricate practical

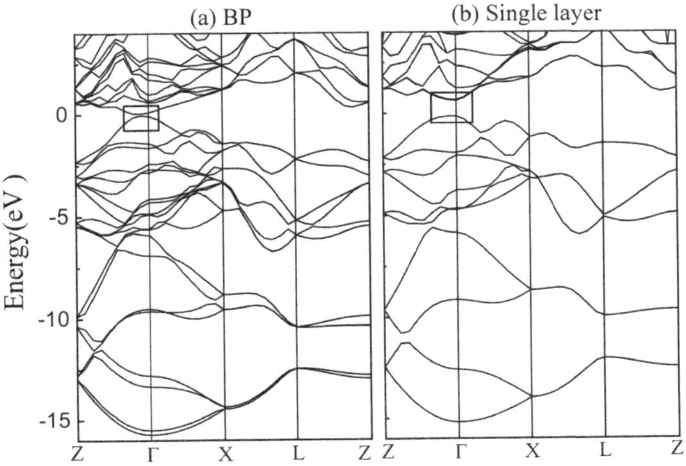

Fig. 3 Electronic band structures of multi-layer and single-layer BP materials; the Fermi-energy has been set to zero [41]

devices based on the few-layer than monolayer nanosheets. As the exciton binding energy (BE) is quite significant in 2D semiconductors, immediate cross-validation of the theoretically estimated electron bandgaps with the experimentally measured optical data may be misleading. While the theoretical values are normally estimated in a vacuo, the empirical data can be collected from samples placed on a dielectric substrate [26].

The high-pressure (10 GPa) phase of P (pure cubic phosphorus) is found to be superconductive [6, 26], with a critical temperature (T_c) that peaks at 9.5–10.7 K at ~30 GPa [26, 43]. The conductive bulk form can act as an electrode surface [40, 44]. The band gap of BP diminishes at higher pressures. Such pressure-dependent changes have only been observed in BP to date [6]. In contrast, ab initio calculations have shown a decrease in T_c with pressure [45]. Hence, the increase in T_c with pressure is yet to be explained fundamentally. A possible explanation for this conflict is that T_c depends on hysteresis in the T–P phase diagram at temperatures lower than 15 K [46], which can be assigned to the simultaneous presence of rhombohedral and black P phases and the resulting crystalline displacements [26]. Even though the BP monolayer is a direct semiconductor, it can be turned into an n-type semiconductor with increased T_c for an electron-doped system [47]. The bilayer BP can also be converted to a Bardeen–Cooper–Schrieffer (BCS)-type (super)conductor through Li doping [26, 48].

4 Optical Properties, Excitons and Photoluminescence

The optical features of BP nanostructures also show high anisotropy. This further leads to linear dichroism, i.e., it can absorb light rays of different polarizations at different rates, particularly at frequencies near that of the bandgap energy [4, 17, 26, 49]. An electronic transition occurs in phosphorene owing to finite dipole matrix elements of $\hat{p}$ between the bottom of the conduction band and the top of the valence band. Phosphorene and few-layer BP absorb only the light rays with a polarization component along the armchair direction, but it becomes transparent to the light polarized in the zigzag direction. Indeed, the transition due to the y-polarized light along the zigzag direction is forbidden by the symmetries of the electron WF in the valence and conduction bands. Therefore, the single-layer and few-layer BP nanostructures are natural optical photon polarizers appropriate for optical devices. This optoelectronic anisotropy can be exploited to indicate the orientation of the crystallographic axes. Furthermore, the bulk BP displays a huge elevation in the optical conductivity owing to the tendency toward band-nesting [49]. Such a behavior renders phosphorene as a promising photodetector with high responsivity in the UV range [17, 26, 42]. The responsivity of the few-layer BP devices is about 100 times higher than the photo-responsivity of graphene [50], but lower than that of TMDs, which falls into the range of 7.5–780 A/W [51]. Nonetheless, TMD-based photodetectors have low mobilities and respond rather slowly to light relative to the phosphorene-based devices [17, 18, 39, 52].

The reduced dimension and lower screening of the Coulomb attraction in monolayer semiconductors give rise to relatively high exciton energies [42]. Polarization-resolved photoluminescence (PL) measurements on phosphorene indicate strongly bound excitons having a BE of 900 ± 100 meV [53]. Such a high value can increase the lifetimes of excitons and trions (charged excitons) under thermal disturbances, thus making these composite particles appropriate for the mobility of photons by trions as well as the optoelectronically controlled quantum computations [26]. The BP monolayer displays charge carrier effective masses heavier along the zigzag direction; this directional mass dependence leads to a strong in-plane anisotropy of excitons in BP with the WF in the armchair direction [3, 26, 53]. Interestingly, the PL intensity has been found to decline exponentially with increasing the layer thickness [17, 54]. Moreover, the thickness of BP nanostructures can be estimated from the size of excitons (Fig. 4) [42]. Unlike some other layered materials, the phosphorene exciton BE (1.31–1.75 eV) is much larger than the bulk BP value (~30 meV as in many semiconductors) [26, 53]. The PL spectra have also detected trions with a BE value of 100 meV in phosphorene supported on a silica material, which is substantially higher than that of TMDCs [26, 55].

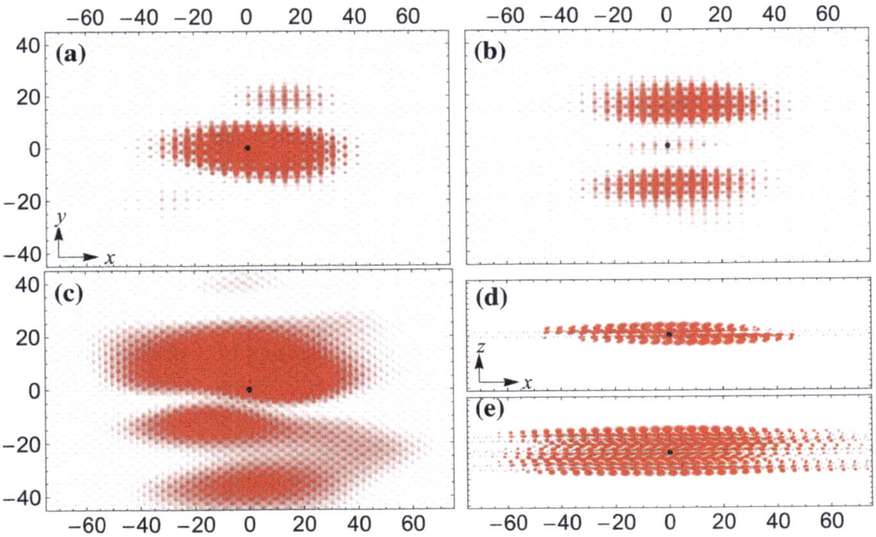

Fig. 4 Top views of the squared electron WFs of the first (**a**) and second (**b**) bound excitons in the single-layer BP, and the first bound exciton in the trilayer BP (**c**). Side views of the WFs in (**a**) and (**c**) are shown in (**d**) and (**e**), respectively, where the hole, depicted by a black dot, is set to the origin and lines illustrating the atomic bonds are superimposed (with the scale in Å) [42]

5 Charge Transport

A remarkable charge carrier mobility (μ_e) found in recent years for BP (particularly in the armchair direction) is an appealing driving force in advancing the preparation and encapsulation methods [26]. Whereas bulk BP shows the electron and hole mobility values of respectively 220 and 350 $cm^2V^{-1}s^{-1}$ at room temperature, the highest hole mobility for the few-layer BP is comparable with that of Si [15]. The carrier mobility was found to increase with the number of layers (up to ca. 10 nm) to values over 1000 $cm^2V^{-1}s^{-1}$ along the *y*-direction. Compared to the fundamental phonon-limited carrier mobility limit, the hole mobility (μ_h) can still amount, e.g., to 4800–6400 $cm^2V^{-1}s^{-1}$ for the BP five-layer flake [17, 26, 56]. Owing to the electron–phonon scattering, however, both hole mobility and Hall mobility drop at higher temperatures (above 100 K) [17].

6 Quantum Oscillations and Magnetism

The observation of quantum effects for a material, here the BP monolayer, serves as an indication of its favored crystal quality to meet specific benchmarks. Fermi energy sweeping takes place through the Landau quantization within an oscillatory resistance in adequately clean materials, such as phosphorene. As the spacing of the adjacent Landau levels are diminutive, low temperatures are needed to correctly identify the mentioned event. For instance, this phenomenon was found along with the existence of a Zeeman splitting at relatively large magnetic fields, namely B > 8T for an encapsulated BP monolayer [57]. Such observations testify the absence of any disorder in the encapsulated sys-tem at hand. Thanks to the high mobility in the few-layer BP (~400 and 4000 $cm^2V^{-1}s^{-1}$ at room-temperature and 1.5 K, respectively) [57], on/off ratios beyond 1,00,000 can be accomplished in BP-based FETs (Fig. 5) [26]. The carrier mobility and the on/off ratio of the resulting FET devices are layer-dependent [3, 6, 15]. Owing to the BP band dispersion anisotropy, the mass associated with the Landau level interspacing is determined as the geometric mean of m_x and m_y; therefore, an ultraclean BP system can be taken as an ideal platform material for studying the peculiarities of mass anisotropy [6, 26]. Anisotropic magnetic properties

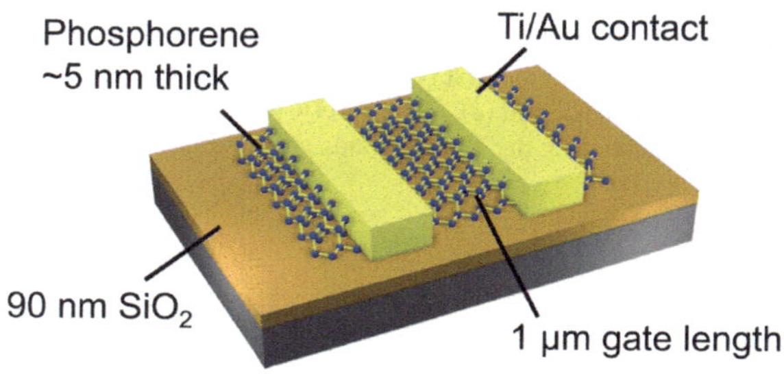

Fig. 5 Schematic diagram of a typical back-gate BP FET [7]

have also been predicted for the BP monolayer [3]. Although black phosphorene is intrinsically non-magnetic, single vacancies can induce local magnetic moments [58, 59]. Notably, a dissipative quantum Hall effect has been estimated for the lattice imperfections (defects) along the zigzag direction [3, 60].

7 Phonons and Plasmons

The BP monolayer shows high anisotropic phonon dispersion curves that give rise to higher phonon group velocity components along the zigzag direction [36]. Substantial anisotropic behavior leads to orientation-dependent thermal transport [61] and shear modulus [3, 62]. While containing four atoms in each unit cell, the BP monolayer has twelve (including nine optical plus three acoustic) phonon modes [26, 63]. The essential phonon branches of phosphorene include B_{2g}, A_{2g} (in-plane), and A_{1g} (out of plane) [3, 64]. The B_{3g} characteristic Raman tensor of the mode has only the yz non-null matrix components. The polarization dependence of Raman scattering thus enables immediate and accurate analysis of the crystalline orientation in the few-layer BP [54]. Nonetheless, one should note that the frequencies of the main high-frequency Raman peaks do not explicitly correlate with the layer number 18 and the band shift is also influenced by some other factors, such as zigzag strain [26, 65]. Unlike an armchair strain, in such a circumstance, B_{2g} and A_{2g} modes will experience blueshift and redshift under compression and stretching, respectively. Instead, the A_{1g} modes undergo redshift in both types of compression and stretching strain [3].

As the anisotropic band dispersion markedly influences the polarization function and consequently the collective excitations, plasmons are also expected to be highly direction-dependent. Even though the complete Hamiltonian for BP is not fairly simple, the plasmonic dispersion may be determined from one anisotropic band structure within the random phase approximation (RPA) [2, 26]. Landau damp-ing refers to a phenomenon of plasmons dissociation via recombining the electrons and holes at the same energy-momentum space. What makes the few-layer BP be prominent among other 2D materials is that the commencement of the Landau damp-ing is quite orientation-dependent. Assuming that a plasmonic branch is located where the band structure forbids such recombination, plasmons can ideally find an infinite lifetime according to RPA. In other words, the anisotropy of the material can be benefited to limit or allow the plasmons propagation [26].

8 Heat Transport and Thermoelectric Properties

Both monolayer and multilayer materials show a significant Seebeck coefficient or thermoelectric sensitivity ($S = 335 \pm 10\ \mathrm{VK}^{-1}$ at room temperature) [66]. Nonetheless, the favored directions of charge and heat conduction in phosphorene are orthogonal. Such a behavior can be an advantage in thermoelectric generators or coolers, which require both low thermal and high electrical conductivity values. As pointed out previously, the effective masses of both 2D holes and electrons in the BP monolayer at the vicinity of the Γ point are significantly higher in the zigzag direction than along the perpendicular one [3, 17, 19, 26].

Furthermore, the phonon-limited electrons and holes are largely aniso-tropic; therefore, the anisotropic conductance of ~10 times larger develops in the armchair direction compared to that in the zigzag one [17, 26, 67]. Instead, the thermal conductivity is larger in the latter direction [26, 67]. The effective masses of charge carriers can also change by strain, where the changes also depend on the strain orientation, serving as further proof of the anisotropic nature of phosphorene [3]. Raman modes also show significant shifts with the uniaxial strains, while the frequency shift depends on the amplitude and direction of strains, the atomic vibrations, and consequently, the changing interatomic distances [3].

Micro-Raman spectroscopy has shown that the thermal conductivity values increase with the BP thickness [68]. The highest figure of merit (ZT) for doped phosphorene has been estimated to be 0.2–0.7 and up to 2.5 at low temperatures and 500 K, respectively [67], and is even higher in nanoribbons [67]. The high-temperature ZT value mentioned above competes with the highest magnitudes reported, e.g., in the Pb chalcogenide alloys [69]. These fantastic properties make phosphorene an excellent thermoelectric device [26].

References

1. Bridgman, P.W.: Two new modifications of phosphorus. J. Am. Chem. Soc. **36**(7), 1344–1363 (1914). https://doi.org/10.1021/ja02184a002
2. Rodin, A.S., Carvalho, A., Castro Neto, A.H.: Strain-induced gap modification in black phosphorus. Phys. Rev. Lett. **112**(17), 176801 (2014). https://doi.org/10.1103/PhysRevLett.112.176801
3. Akhtar, M., Anderson, G., Zhao, R., Alruqi, A., Mroczkowska, J.E., Sumanasekera, G., Jasinski, J.B.: Recent advances in synthesis, properties, and applications of phosphorene. npj 2D Mater. Appl. **1**(1):5 (2017). https://doi.org/10.1038/s41699-017-0007-5
4. Du, Y., Luo, Z., Liu, H., Xu, X., Ye, P.D.: Anisotropic properties of black phosphorus. In: Avouris, P., Low, T., Heinz, T.F. (eds.) 2D Materials: Properties and Devices, pp. 413–434. Cambridge University Press, Cambridge (2017). https://doi.org/10.1017/9781316681619.023
5. Jing, Y., Zhang, X., Zhou, Z.: Phosphorene: what can we know from computations? WIREs Comput. Mol. Sci. **6**(1), 5–19 (2016). https://doi.org/10.1002/wcms.1234
6. Liu, H., Neal, A.T., Zhu, Z., Luo, Z., Xu, X., Tománek, D., Ye, P.D.: Phosphorene: an unexplored 2D semiconductor with a high hole mobility. ACS Nano **8**(4), 4033–4041 (2014). https://doi.org/10.1021/nn501226z

7. Liu, H., Du, Y., Deng, Y., Ye, P.D.: Semiconducting black phosphorus: synthesis, transport properties and electronic applications. Chem. Soc. Rev. **44**(9), 2732–2743 (2015). https://doi.org/10.1039/c4cs00257a
8. Ghashghaee, M., Ghambarian, M.: Adsorption of toxic mercury, lead, cadmium, and arsenic ions on black phosphorous nanosheet: first-principles calculations. Struct. Chem. **30**(1), 85–96 (2019). https://doi.org/10.1007/s11224-018-1173-6
9. Hirsch, A., Hauke, F.: Post-graphene 2D chemistry: the emerging field of molybdenum disulfide and black phosphorus functionalization. Angew. Chem. Int. Ed. **57**(16), 4338–4354 (2018). https://doi.org/10.1002/anie.201708211
10. Abellán, G., Wild, S., Lloret, V., Scheuschner, N., Gillen, R., Mundloch, U., Maultzsch, J., Varela, M., Hauke, F., Hirsch, A.: Fundamental insights into the degradation and stabilization of thin layer black phosphorus. J. Am. Chem. Soc. **139**(30), 10432–10440 (2017). https://doi.org/10.1021/jacs.7b04971
11. Grayfer, E.D., Kozlova, M.N., Fedorov, V.E.: Colloidal 2D nanosheets of MoS_2 and other transition metal dichalcogenides through liquid-phase exfoliation. Adv. Coll. Interface Sci. **245**, 40–61 (2017). https://doi.org/10.1016/j.cis.2017.04.014
12. Yasaei, P., Kumar, B., Foroozan, T., Wang, C., Asadi, M., Tuschel, D., Indacochea, J.E., Klie, R.F., Salehi-Khojin, A.: High-quality black phosphorus atomic layers by liquid-phase exfoliation. Adv. Mater. **27**(11), 1887–1892 (2015). https://doi.org/10.1002/adma.201405150
13. Favron, A., Gaufrès, E., Fossard, F., Phaneuf-L'Heureux, A.-L., Tang, N.Y.W., Lévesque, P.L., Loiseau, A., Leonelli, R., Francoeur, S., Martel, R.: Photooxidation and quantum confinement effects in exfoliated black phosphorus. Nat. Mater. **14**, 826 (2015). https://doi.org/10.1038/nmat4299
14. Abellán, G., Lloret, V., Mundloch, U., Marcia, M., Neiss, C., Görling, A., Varela, M., Hauke, F., Hirsch, A.: Noncovalent functionalization of black phosphorus. Angew. Chem. Int. Ed. **55**(47), 14557–14562 (2016). https://doi.org/10.1002/anie.201604784
15. Lei, W., Liu, G., Zhang, J., Liu, M.: Black phosphorus nanostructures: recent advances in hybridization, doping and functionalization. Chem. Soc. Rev. **46**(12), 3492–3509 (2017). https://doi.org/10.1039/c7cs00021a
16. Hultgren, R., Gingrich, N.S., Warren, B.E.: The atomic distribution in red and black phosphorus and the crystal structure of black phosphorus. J. Chem. Phys. **3**(6), 351–355 (1935). https://doi.org/10.1063/1.1749671
17. Sorkin, V., Cai, Y., Ong, Z., Zhang, G., Zhang, Y.W.: Recent advances in the study of phosphorene and its nanostructures. Crit. Rev. Solid State Mater. Sci. **42**(1), 1–82 (2017). https://doi.org/10.1080/10408436.2016.1182469
18. Xia, F., Wang, H., Jia, Y.: Rediscovering black phosphorus as an anisotropic layered material for optoelectronics and electronics. Nat. Commun. **5**, 4458 (2014). https://doi.org/10.1038/ncomms5458
19. Chowdhury, C., Datta, A.: Exotic physics and chemistry of two-dimensional phosphorus: phosphorene. J. Phys. Chem. Lett. **8**(13), 2909–2916 (2017). https://doi.org/10.1021/acs.jpclett.7b01290
20. Samuel Reich, E.: Phosphorene excites materials scientists. Nat. News **506**(7486), 19 (2014)
21. Morita, A.: Semiconducting black phosphorus. Appl. Phys. A **39**(4), 227–242 (1986). https://doi.org/10.1007/bf00617267
22. Shulenburger, L., Baczewski, A.D., Zhu, Z., Guan, J., Tománek, D.: The nature of the interlayer interaction in bulk and few-layer phosphorus. Nano Lett. **15**(12), 8170–8175 (2015). https://doi.org/10.1021/acs.nanolett.5b03615
23. Cai, Y., Zhang, G., Zhang, Y.-W.: Layer-dependent band alignment and work function of few-layer phosphorene. Sci. Rep. **4**, 6677 (2014). https://doi.org/10.1038/srep06677
24. Graziano, G., Klimeš, J., Fernandez-Alonso, F., Michaelides, A.: Improved description of soft layered materials with van der Waals density functional theory. J. Phys.: Condens. Matter **24**(42), 424216 (2012). https://doi.org/10.1088/0953-8984/24/42/424216
25. Hart, R.R., Robin, M.B., Kuebler, N.A.: 3p orbitals, bent bonds, and the electronic spectrum of the P_4 molecule. J. Chem. Phys. **42**(10), 3631–3638 (1965). https://doi.org/10.1063/1.1695771

26. Carvalho, A., Wang, M., Zhu, X., Rodin, A.S., Su, H., Castro Neto, A.H.: Phosphorene: from theory to applications. Nat. Rev. Mater. **1**, 16061 (2016). https://doi.org/10.1038/natrevmats.2016.61
27. Castellanos-Gomez, A., Vicarelli, L., Prada, E., Island, J.O., Narasimha-Acharya, K.L., Blanter, S.I., Groenendijk, D.J., Buscema, M., Steele, G.A., Alvarez, J.V., Zandbergen, H.W., Palacios, J.J., van der Zant, H.S.J.: Isolation and characterization of few-layer black phosphorus. 2D Mater. **1**(2), 025001 (2014). https://doi.org/10.1088/2053-1583/1/2/025001
28. Appalakondaiah, S., Vaitheeswaran, G., Lebègue, S., Christensen, N.E., Svane, A.: Effect of van der Waals interactions on the structural and elastic properties of black phosphorus. Phys. Rev. B **86**(3), 035105 (2012). https://doi.org/10.1103/PhysRevB.86.035105
29. Lam, K., Dong, Z., Guo, J.: Performance limits projection of black phosphorous field-effect transistors. IEEE Electron Dev. Lett. **35**(9), 963–965 (2014). https://doi.org/10.1109/led.2014.2333368
30. Hu, Z.-X., Kong, X., Qiao, J., Normand, B., Ji, W.: Interlayer electronic hybridization leads to exceptional thickness-dependent vibrational properties in few-layer black phosphorus. Nanoscale **8**(5), 2740–2750 (2016). https://doi.org/10.1039/c5nr06293d
31. Lui, C.H., Ye, Z., Keiser, C., Xiao, X., He, R.: Temperature-activated layer-breathing vibrations in few-layer graphene. Nano Lett. **14**(8), 4615–4621 (2014). https://doi.org/10.1021/nl501678j
32. Low, T., Rodin, A.S., Carvalho, A., Jiang, Y., Wang, H., Xia, F., Castro Neto, A.H.: Tunable optical properties of multilayer black phosphorus thin films. Phys. Rev. B **90**(7), 075434 (2014)
33. Cartz, L., Srinivasa, S.R., Riedner, R.J., Jorgensen, J.D., Worlton, T.G.: Effect of pressure on bonding in black phosphorus. J. Chem. Phys. **71**(4), 1718–1721 (1979). https://doi.org/10.1063/1.438523
34. Jiang, J.-W., Park, H.S.: Negative poisson's ratio in single-layer black phosphorus. Nat. Commun. **5**, 4727 (2014). https://doi.org/10.1038/ncomms5727
35. Zhang, S., Guo, S., Chen, Z., Wang, Y., Gao, H., Gómez-Herrero, J., Ares, P., Zamora, F., Zhu, Z., Zeng, H.: Recent progress in 2D group-VA semiconductors: from theory to experiment. Chem. Soc. Rev. **47**(3), 982–1021 (2018). https://doi.org/10.1039/c7cs00125h
36. Jain, A., McGaughey, A.J.H.: Strongly anisotropic in-plane thermal transport in single-layer black phosphorene. Sci. Rep. **5**, 8501 (2015). https://doi.org/10.1038/srep08501
37. Zhu, Z., Tománek, D.: Semiconducting layered blue phosphorus: a computational study. Phys. Rev. Lett. **112**(17), 176802 (2014). https://doi.org/10.1103/PhysRevLett.112.176802
38. Guan, J., Zhu, Z., Tománek, D.: Phase coexistence and metal-insulator transition in few-layer phosphorene: a computational study. Phys. Rev. Lett. **113**(4), 046804 (2014). https://doi.org/10.1103/PhysRevLett.113.046804
39. Dai, J., Zeng, X.C.: Bilayer phosphorene: effect of stacking order on bandgap and its potential applications in thin-film solar cells. J. Phys. Chem. Lett. **5**(7), 1289–1293 (2014). https://doi.org/10.1021/jz500409m
40. Pumera, M.: Phosphorene and black phosphorus for sensing and biosensing. TrAC Trends Anal. Chem. **93**, 1–6 (2017). https://doi.org/10.1016/j.trac.2017.05.002
41. Du, Y., Ouyang, C., Shi, S., Lei, M.: Ab initio studies on atomic and electronic structures of black phosphorus. J. Appl. Phys. **107**(9), 093718 (2010). https://doi.org/10.1063/1.3386509
42. Tran, V., Soklaski, R., Liang, Y., Yang, L.: Layer-controlled band gap and anisotropic excitons in few-layer black phosphorus. Phys. Rev. B **89**(23), 235319 (2014). https://doi.org/10.1103/PhysRevB.89.235319
43. Karuzawa, M., Ishizuka, M., Endo, S.: The pressure effect on the superconducting transition temperature of black phosphorus. J. Phys.: Condens. Matter **14**(44), 10759–10762 (2002). https://doi.org/10.1088/0953-8984/14/44/372
44. Wang, L., Sofer, Z., Pumera, M.: Voltammetry of layered black phosphorus: electrochemistry of multilayer phosphorene. ChemElectroChem **2**(3), 324–327 (2015). https://doi.org/10.1002/celc.201402363
45. Chan, K.T., Malone, B.D., Cohen, M.L.: Pressure dependence of superconductivity in simple cubic phosphorus. Phys. Rev. B **88**(6), 064517 (2013). https://doi.org/10.1103/PhysRevB.88.064517

46. Kawamura, H., Shirotani, I., Tachikawa, K.: Anomalous superconductivity in black phosphorus under high pressures. Solid State Commun. **49**(9), 879–881 (1984). https://doi.org/10.1016/0038-1098(84)90444-7
47. Shao, D.F., Lu, W.J., Lv, H.Y., Sun, Y.P.: Electron-doped phosphorene: a potential monolayer superconductor. EPL (Europhys. Lett.) **108**(6), 67004 (2014). https://doi.org/10.1209/0295-5075/108/67004
48. Huang, G.Q., Xing, Z.W., Xing, D.Y.: Prediction of superconductivity in Li-intercalated bilayer phosphorene. Appl. Phys. Lett. **106**(11), 113107 (2015). https://doi.org/10.1063/1.4916100
49. Yuan, H., Liu, X., Afshinmanesh, F., Li, W., Xu, G., Sun, J., Lian, B., Curto, A.G., Ye, G., Hikita, Y., Shen, Z., Zhang, S.-C., Chen, X., Brongersma, M., Hwang, H.Y., Cui, Y.: Polarization-sensitive broadband photodetector using a black phosphorus vertical p–n junction. Nat. Nanotechnol. **10**, 707 (2015). https://doi.org/10.1038/nnano.2015.112
50. Low, T., Engel, M., Steiner, M., Avouris, P.: Origin of photoresponse in black phosphorus phototransistors. Phys. Rev. B **90**(8), 081408 (2014). https://doi.org/10.1103/PhysRevB.90.081408
51. Lopez-Sanchez, O., Lembke, D., Kayci, M., Radenovic, A., Kis, A.: Ultrasensitive photodetectors based on monolayer MoS_2. Nat. Nanotechnol. **8**, 497 (2013). https://doi.org/10.1038/nnano.2013.100
52. Deng, Y., Luo, Z., Conrad, N.J., Liu, H., Gong, Y., Najmaei, S., Ajayan, P.M., Lou, J., Xu, X., Ye, P.D.: Black phosphorus-monolayer MoS_2 van der Waals heterojunction p–n diode. ACS Nano **8**(8), 8292–8299 (2014). https://doi.org/10.1021/nn5027388
53. Wang, X., Jones, A.M., Seyler, K.L., Tran, V., Jia, Y., Zhao, H., Wang, H., Yang, L., Xu, X., Xia, F.: Highly anisotropic and robust excitons in monolayer black phosphorus. Nat. Nanotechnol. **10**, 517 (2015). https://doi.org/10.1038/nnano.2015.71
54. Zhang, S., Yang, J., Xu, R., Wang, F., Li, W., Ghufran, M., Zhang, Y.-W., Yu, Z., Zhang, G., Qin, Q., Lu, Y.: Extraordinary photoluminescence and strong temperature/angle-dependent raman responses in few-layer phosphorene. ACS Nano **8**(9), 9590–9596 (2014). https://doi.org/10.1021/nn503893j
55. Yang, J., Xu, R., Pei, J., Myint, Y.W., Wang, F., Wang, Z., Zhang, S., Yu, Z., Lu, Y.: Optical tuning of exciton and trion emissions in monolayer phosphorene. Light: Sci. Appl. **4**, e312 (2015). https://doi.org/10.1038/lsa.2015.85
56. Qiao, J., Kong, X., Hu, Z.-X., Yang, F., Ji, W.: High-mobility transport anisotropy and linear dichroism in few-layer black phosphorus. Nat. Commun. **5**, 4475 (2014). https://doi.org/10.1038/ncomms5475
57. Gillgren, N., Wickramaratne, D., Shi, Y., Espiritu, T., Yang, J., Hu, J., Wei, J., Liu, X., Mao, Z., Watanabe, K., Taniguchi, T., Bockrath, M., Barlas, Y., Lake, R.K., Ning Lau, C.: Gate tunable quantum oscillations in air-stable and high mobility few-layer phosphorene heterostructures. 2D Mater. **2**(1):011001 (2014). https://doi.org/10.1088/2053-1583/2/1/011001
58. Srivastava, P., Hembram, K.P.S.S., Mizuseki, H., Lee, K.-R., Han, S.S., Kim, S.: Tuning the electronic and magnetic properties of phosphorene by vacancies and adatoms. J. Phys. Chem. C **119**(12), 6530–6538 (2015). https://doi.org/10.1021/jp5110938
59. Ren, J., Zhang, C., Li, J., Guo, Z., Xiao, H., Zhong, J.: Strain engineering of magnetic state in vacancy-doped phosphorene. Phys. Lett. A **380**(40), 3270–3277 (2016). https://doi.org/10.1016/j.physleta.2016.07.055
60. Ostahie, B., Aldea, A.: Phosphorene confined systems in magnetic field, quantum transport, and superradiance in the quasiflat band. Phys. Rev. B **93**(7), 075408 (2016). https://doi.org/10.1103/PhysRevB.93.075408
61. Jang, H., Wood, J.D., Ryder, C.R., Hersam, M.C., Cahill, D.G.: Anisotropic thermal conductivity of exfoliated black phosphorus. Adv. Mater. **27**(48), 8017–8022 (2015). https://doi.org/10.1002/adma.201503466
62. Cai, Y., Ke, Q., Zhang, G., Feng, Y.P., Shenoy, V.B., Zhang, Y.-W.: Giant phononic anisotropy and unusual anharmonicity of phosphorene: interlayer coupling and strain engineering. Adv. Func. Mater. **25**(15), 2230–2236 (2015). https://doi.org/10.1002/adfm.201404294

63. Ribeiro-Soares, J., Almeida, R.M., Cançado, L.G., Dresselhaus, M.S., Jorio, A.: Group theory for structural analysis and lattice vibrations in phosphorene systems. Phys. Rev. B **91**(20), 205421 (2015). https://doi.org/10.1103/PhysRevB.91.205421
64. Fei, R., Yang, L.: Lattice vibrational modes and Raman scattering spectra of strained phosphorene. Appl. Phys. Lett. **105**(8), 083120 (2014). https://doi.org/10.1063/1.4894273
65. Fei, R., Yang, L.: Strain-engineering the anisotropic electrical conductance of few-layer black phosphorus. Nano Lett. **14**(5), 2884–2889 (2014). https://doi.org/10.1021/nl500935z
66. Flores, E., Ares, J.R., Castellanos-Gomez, A., Barawi, M., Ferrer, I.J., Sánchez, C.: Thermoelectric power of bulk black-phosphorus. Appl. Phys. Lett. **106**(2), 022102 (2015). https://doi.org/10.1063/1.4905636
67. Fei, R., Faghaninia, A., Soklaski, R., Yan, J.-A., Lo, C., Yang, L.: Enhanced thermoelectric efficiency via orthogonal electrical and thermal conductances in phosphorene. Nano Lett. **14**(11), 6393–6399 (2014). https://doi.org/10.1021/nl502865s
68. Luo, Z., Maassen, J., Deng, Y., Du, Y., Garrelts, R.P., Lundstrom, M.S., Ye, P.D., Xu, X.: Anisotropic in-plane thermal conductivity observed in few-layer black phosphorus. Nat. Commun. **6**, 8572 (2015). https://doi.org/10.1038/ncomms9572
69. Wu, H.J., Zhao, L.D., Zheng, F.S., Wu, D., Pei, Y.L., Tong, X., Kanatzidis, M.G., He, J.Q.: Broad temperature plateau for thermoelectric figure of merit ZT > 2 in phase-separated $PbTe_{0.7}S_{0.3}$. Nat. Commun. **5**, 4515 (2014). https://doi.org/10.1038/ncomms5515

Black Phosphorous Quantum Dots

S. Anju, N. Prajitha, V. G. Reshma and P. V. Mohanan

Abstract Black phosphorus a.k.a phosphorene is a very recent offshoot in the genre of 2D materials, which gained immense scientific engrossment ever since its appreciable rediscovery in 2014. Ascribe to the outstanding inherent characteristics like tunable bandgap, good enough carrier mobility, size tunable absorption spectra, narrow emission spectra, potent biodegradability and excellent biocompatibility; BP has become an encouraging successor for various applications in biomedical sector. In the zero dimensional state, black phosphorus quantum dots (BPQDs) has become a crucial substitute for various currently employed carbon based and semiconductor quantum dots. Apart from all the exhibited physico-chemical characteristics of its 2D form, BPQDs has contributed more towards optical applications as well. Even though the outlook seems to be extremely well adapted for biomedicine, practical applications are still highly challenging especially due to its fast degradation and instability under ambient conditions of air and moisture. Therefore proper characterisation and surface modifications should be done to unveil these challenges. The present chapter has adopted strategies to give precise attention inorder familiarise upon the various physico-chemical properties, synthesis methods as well as stability of BPQDs under ambient conditions. Also focussed on the various biomedical applications of BPQDs along with briefly underlining the applications of BPQDs in other research arenas like optoelectronics, photo catalysis etc., Briefly discussed about the possible toxic impacts of this emerging material also.

Keywords Black phosphorous · Quantum dots · Biomedical · Drug delivery · Imaging

All Authors contributed equally.

S. Anju · N. Prajitha · V. G. Reshma · P. V. Mohanan (✉)
Toxicology Division, Biomedical Technology Wing, Sree Chitra Tirunal Institute for Medical Sciences and Technology, Thiruvananthapuram 695 012, Kerala, India
e-mail: mohanpv10@gmail.com; mohanpv@sctimst.ac.in

Inamuddin et al. (eds.), *Black Phosphorus*, Engineering Materials,
https://doi.org/10.1007/978-3-030-29555-4_4

1 Introduction

Black Phosphorous Quantum Dots (BPQDs) are zero dimensional functional nanomaterials emerged as quick highflier on the horizons of materials science and chemistry. Black phosphorus (BP) nanostructures mainly exist in two forms, one is two-dimensional (2D) BP layered nanosheets and one-dimensional (1D) BP nanoribbons. 0D BPQDs is another form of black phosphorus (BP) nanostructures that were first synthesized in 2015 by Zhang et al., through a top-down method. BPQDs maintain an exclusively refined structural arrangement similar to a puckered network of connected hinges. BP exist in two basic structures. One is the puckered structure along the armchair direction parallel to the x-axis and the latter one is the bilayer configuration along the zigzag direction parallel to the y-axis. BPQDs display good ultraviolet visible absorption spectroscopy. These QDs also have good photo stability with large extinction coefficient of 14.8 Lg^{-1} cm^{-1} at 808 nm. It has been reported that the band gap of BP can arrive at approximately 1.45 eV. BPQDs are formed by reducing size of the black phosphorus to nanoscale dimension [1]. It possess wide band gap than layered black phosphorus due to the renowned QD confinement effect. BPQDs have received considerable attention because of their exceptional properties applicable for vapour sensors [2, 3], electronic devices [4] and cancer therapy [5, 6]. BPQDs are capable tool for various tumor ablation therapies due to its high near-infrared (NIR) extinction coefficient and photothermal efficiency [7]. BPQDs have been shown to have powerful absorption of solar light and larger surface area which brings about possibilities capable for photocatalysis [8].

2 History of Discovery

Black phosphorous is a new member of the 2D material family. It has very recently started to gain immense attention due of its high carrier mobility, finite band gap, and large ON-OFF current ratios. Phosphorus is one of the indispensable elements of nucleic acids and bone mass so that it plays crucial role for maintaining major physiological processes in the human body. 1% of total body mass approximately 660 g is purely composed of mineral phosphorus. It is a highly biocompatible material with extensive application in biomedical technology. It has three major isotopes, including white phosphorus (WP), red phosphorus (RP) and BP. Among these three, the most stable form is black phosphorus and it is inflammable under any extreme conditions and insoluble in almost all solvents in the bare form. WP is a white-yellow transparent solid which possess a tetrahedral structure. It is chemically very much unstable and highly reactive due to large bond strain. RP possess an amorphous structure. It shows high electrical resistivity and good enough chemical stability when compared to WP [9].

BP was first discovered by Bridgman in [10]. Later then it did not gain much attention over the last century, because of its insensitive preparation conditions and little knowledge about possible applications. Later on after 2014, monolayer phosphorene has been exfoliated mechanically from its bulk counterpart, which was found to be connected by weak van der Waals forces between stacked layers. Earlier reports have described the preparation of BP nanostructures from liquid exfoliation of bulk BP crystals besides mechanical exfoliation. 2D BP layered structures have layer dependent band gap that can be tuned from 0.3 eV (bulk) to 2.0 eV (single layer). Due to this property, BP nanostructures display great promise in many significant areas, including lithium ion batteries, field-effect transistors, memory devices, demodulators, photodetectors and diodes. Ascribe to stringent electronic features exhibited by 1D BP nanoribbons, they got scientifically explored to a considerable extend, which later on found to be highly susceptible to various imperfections due to edge-structure, edge-width and edge-type defects. Most recently, another BP nanostructure, zero-dimensional (0D) BP quantum dots (QDs), was synthesized by various chemical methods. Zhang et al. [4] has successfully synthesized BPQDs for the first time. Because of the exceptional physicochemical properties, BPQDs cover a broad range of fields, especially bioimaging, nonlinear optical absorbers, fluorescence sensing, photothermal cancer therapy, intelligent electronic elements, optoelectronics, photovoltaics and flexible devices.

3 Properties

BPQDs have been exposed to have stronger absorption of entire UV/Visible/NIR spectral range and larger surface areas which makes it more efficient candidate for photocatalysis [8]. However, under visible light irradiation, BP get oxidized and degraded into PxOy which is sensitive to H_2O and O_2. BPQDs have wide range of applications in various fields because they exhibit many important properties. These properties are mainly because of its unique structural features. Unique properties of BPQDs, including their energy spectral states, electronic and optical properties, are of interest in widespread research fields.

3.1 Energy Spectra and Energy States

Band gap of BPQDs are having deep interface states. BPQDs are not perfectly circular and have mixed edges. As a result band width varies with changes in the size of the QDs. In [11], the charging energy spectra of BPQDs investigated by Lino et al. They achieved a theoretical evaluation of the charging effects in single or double layered BPQDs with lateral sizes of 2 and 3 nm. Charging of BPQDs was influenced by the lateral size and layer number of QDs. These QDs can store up to Nmax electrons during their structural instability. On comparison of the average energy (EA) of similar sized QDs, single-layered one is higher than that of double-layered QDs. The charging energy of BPQDs is dependent on the substrate. Sousa et al. [12]

investigated the influence of substrates such as isolated Si and SiO_2 in the exciton gaps and spectra of the QDs. They also gave a theoretical idea about the significance of dielectric constant of substrate in predicting the excitonic properties of BPQDs. Si and SiO_2 are the substrates with different week and strong dielectric constants.

3.2 Optical Properties of BPQDs

BPQDs exhibit a wide range of absorption spectra that spans over the entire UV/Visible region owing to tunable size dependant characteristics. Niu et al. [1] studied influence of size in determining absorption and emission of BPQDs. They concluded the fact that absorption gap is inversely proportional to the QDs diameter. Although the emission wavelength keep an eye on a mixed behaviour BPQDs exhibits a narrow size dependant emission spectra. It was reported that the blue-shifted emission will be seen if the size increases within the range of 0.8–1.8 nm and is red shifted above 1.8 nm [1]. This anomalism arises from the structural alterations induced by the perfect quantum confinement effect and huge Stokes shift exhibited by these nanostructures. This typical emission performance of BPQDs is entirely discrepant from those of already reported and extensively studied QDs such as PbSe, CdSe, CdSeTe, ZnO, graphene etc, which exhibits a red-shifted emission spectra with increase of size. Small silicon clusters of size range 1.5 nm also shows the same effect [13–15].

3.3 Electronic Properties

Electronic properties of BPQDs depend on their size as well as thickness. By changing their size and thickness, structure and band energy levels can be tuned. BPQDs are inevitable additive for organic photovoltaic devices (OPVs). The properties like their 2D structure as well as strong broadband light absorption and scattering enhanced light absorption contributes to this stringent feature. BPQDs were utilized in perovskite solar cells for the enhancement of hole extraction at the anode side [16].

4 Methods of Synthesis

Zhang et al. [4] first reported the synthesis of BPQDs from bulk BP crystals using a facile top-down (solution phase) method (Fig. 1). Various BPQDs synthetic methods currently employed are pulsed laser irradiation, blender breaking, electrochemical exfoliation, milling crushing and solvothermal treatment.

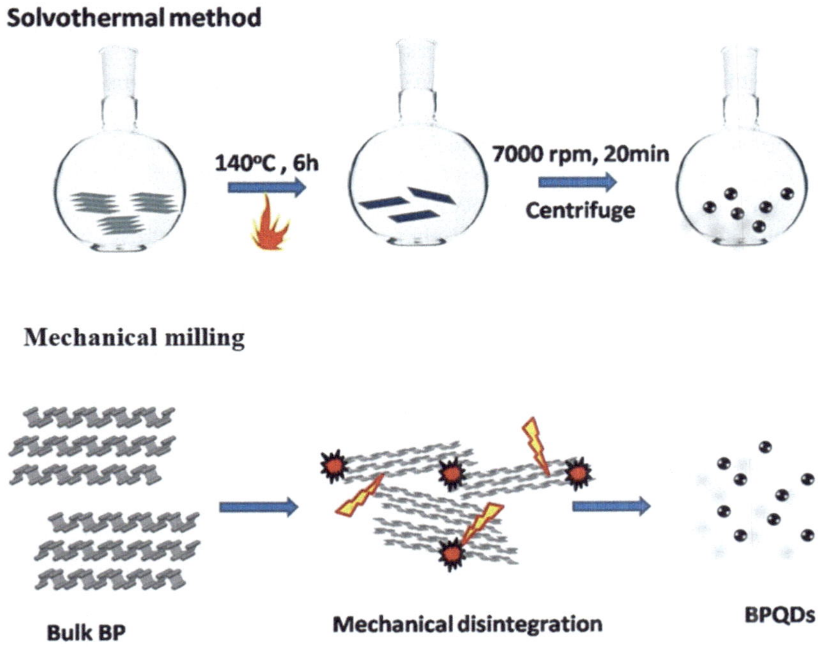

Fig. 1 Synthesis of black phosphorus QDs via solvothermal and mechanical milling method

4.1 Ultrasonic Exfoliation

Ultrasonic exfoliation is a highly proficient facile solvent based method for preparing BPQDs. First attempt of BPQDs synthesis by means of liquid-phase ultrasonic exfoliation is done by Zhang et al. [4]. BP powder and N-methyl pyrrolidone (NMP) were mixed and milled in a mortar and the corresponding admixture was sonicated with an output power of 200 W in an ice bath for 3 h. This was then subjected to 20 min centrifugation at 7000 rpm. BPQDs present in the supernatant was separated. Normally, liquid exfoliation is used to isolate 2D layered materials and formulate QDs. Main disadvantage of this method is wide size distribution and relatively low yield. Consequently, researchers integrated the grinding and sonication processes to exfoliate bulk BP crystals in solution phase, promoting the formation of high-yield BPQDs. BPQDs prepared by two-step ultrasonic strategy by Sun et al. [5] using BP powder as the precursor and NMP as the solvent. Exfoliations of 2D layered materials are commonly forming by probe and bath sonication methods. BP powder was sonicated by probes into BP nanosheets, followed by ice-bath sonication into BPQDs. Water-bath sonication adequate than probe sonication in crushing bulk BP into nanosheets and nanodots But use of bath sonication alone may result in irregular BP nanosheets. Probe sonication is more reproducible than grinding BP crystals into powders. Thus grouping of probe sonication and bath sonication enables the exfoliation of bulk BP crystals into ultra-small BPQDs. Ice-bath sonication or low-temperature cooling is required in probe sonication due to the copious release of heat [5].

Ultrasonic exfoliation is the widely used method of BPQDs preparation. This method avoids all the draw backs of the above detailed method. Organic solvents like NMP, N-vinyl pyrrolidone (NVP) Liu et al. [17], isopropanol (IPA) [18], and dimethyl formamide (DMF) [19]. Higher sonication power (1200 W) and extended sonication time is used in this method. This will helps in achieving small size and promote monodispersity of the BPQDs, which may be critical for enlightening their physicochemical properties and multifunctional applications. BPQDs produced by this method have 4.9 nm diameters, with approximately 5 and 3 layers.

4.2 Solvothermal Treatment

BPQDs can be prepared by solvothermal methods using solvent assisted temperature treatment. This method can be used to produce ultra-small BPQDs on a large scale when compared to mechanical and liquid-phase exfoliation techniques. Bulk BP was dispersed in NMP and maintained for 12 h at 140 °C under vigorous stirring and N_2 atmosphere. The light yellow reaction suspension was centrifuged to separate the unexfoliated large BP crystals. The thickness of the BPQDs was found to be 1.2 nm with a bilayer structure [20].

4.3 Blender Breaking

Blender breaking is a rapid, facile and efficient synthetic method. Zhu et al. [3] reported a new strategy for an ultrablend top-down synthesis of BPQDs using a household kitchen blender. Mechanism behind the formation of BPQDs is the highly turbulent shear rate which will lead to layer by layer disintegration of bulk BP crystals. Bulk BP crystals were dispersed in dimethyl sulphoxide (DMSO) and treated on a household kitchen blender. When the blender start stirring, the BP crystals were found to be swiftly move up from the bottom of the vessel. DMSO has the capacity to disrupt the van der Waals forces exist between the BP interlayers. When the BP crystals were exfoliated and disintegrated under a high-shear turbulence flow, the colour of the dispersion changed to brown and the BPQDs dispersion eventually formed. Colour of the dispersion changed to brown when the BP crystals were exfoliated and disintegrated under a high-shear turbulence flow, which eventually forms a stable BPQD dispersion. Large quantities of uniform BPQDs appeared, indicating the feasibility of preparing BPQDs using a kitchen blender. The average size of the BPQDs synthesised as per the above technique was found to be approximately

2.25 nm. The BPQDs have uniform, well-crystallized and ultra-small sizes structures by the scalable, ultrafast and efficient kitchen blender method [21].

4.4 Milling Crushing

BP crystals were milled in an agate mortar and ultrasonicated in suitable solvent, followed by mechanical milling for 30 min [2]. Depending on the milling period, speed and solvent, the average diameters of the BP NPs varies. Small sized BPQDs were obtained under high-speed centrifugation, that formed stable colloidal suspensions in DMF and NMP.

4.5 Pulsed Laser Irradiation

Pulsed laser irradiation technique produces BPQDs of having outstanding photo luminescent characteristics. When compared to ultrasonic exfoliation, this technique employs a high yield synthesis of photoluminescence (PL) showing BPQDs. Pulsed laser method of intercalation of solid materials in liquids is an effortless technique for the preparation of metal NPs, semiconductor NPs and carbon-related nanomaterials. For example, Ge et al. [22] described in detail about few layer phosphorene QDs acquired by pulsed laser irradiation of bulk BP in diethyl ether.

5 Stability and Surface Modification

Although black phosphorus seems to be a prospective material for various biomedical applications, one factor that hinders its entreaty is lack of stability when exposed to ambient environmental conditions. Especially with air, light and water molecule exposure, the chemical as well as physical parameters gets severely affected leading to fast degradation. Studies have reported that long term exposure can completely vanish away this chemical moiety. Even though this rapid degradation suggests potential for in vivo cancer therapy, this is one major drawback of BP while considering its optical sensing and imaging applications. Concerning the oxygen incorporated degradation of BP; there are two major routes of oxygen incorporation onto BP surface. First one, oxygen molecules can spread through a perpendicular conformation along the surface of BP and gets converted into an oxidised form known as chemisorbed phosphorene [23]. Another possibility is the incorporation of interstitial oxygen that greatly hampers with its structural organisation. Both these chemisorbed and interstitial oxygen molecules will affect its anisotropic properties making the material highly impure [24]. Oxygen atom incorporation greatly increases the hydrophilicity of BP surface making more and more water molecules occupy the exposed surface layer. Photo assisted oxidation reactions favoured by oxygen molecules is also another possibility [25]. Thus oxygen plays an initiator role for various factors of BP degradation. Due to interstitial oxygen atom incorporation, mainly two types of conformational defects can be possible namely diagonal bridge conformation and horizontal bridge conformation. The first possibility mainly concerned with, oxygen molecule incorporation amidst two phosphorus atoms at different layers, which is mainly observed in 2D nanosheet structure and the latter one where oxygen atoms establish association with equidistantly located adjacent phosphorus atoms. Both of which greatly imparts huge structural defects to the intact BP structure [24]. Especially bridge conformation cause huge donor-acceptor depression sites in the interstitial space between phosphorus atoms making a good enough reactive space, which cause huge structural defects when compared to chemisorbed oxygen molecules. Light and temperature also plays crucial role in BP degradation but the exact mechanism is still far from clear cut picture. A recent study reported the influence of discrete UV blue light emission wavelength on BP degradation and the results obtained using sophisticated instrumentation and characterisation proved that ultraviolet spectral component is responsible for BP degradation under ambient light exposure [26].

Several reports suggest that water alone will not lead to extreme deterioration of BP whereas in presence of aerated conditions, water molecule based degradation happens quickly [27]. Also there is no evidence about BP degradation with air and water exposure under dark atmosphere. Therefore to avoid degradation, pristine BP molecules should be treated under perfect darkness and vacuum conditions strictly avoiding oxygen involvement. To overcome this instability, surface functionalization and passivation using various other chemical moieties can be done. Several studies have reported the possibilities of this approach. Surface capping using various 2D materials like graphene, hexagonal boron nitride (h-BN), aluminium oxide and graphitic carbon nitride could effectively improve BP stability under degradation conditions [28]. Another study reported that TiO_2 nanosheet functionalised BP nanostructures revealed considerably long stability under hydrogen-oxygen environment and the results showed that almost 92% of BP composite remained intact after three cycles of 120 min each UV-Vis light irradiation [29]. Surface passivation using aluminium oxide layers and aryl diazonium covalent modification also proved to improve the strategy to a greater extend especially for FET biosensor applications. Specially controlled potassium doping over BP surface is another strategy that can be adopted for the same. PEG, PMMA, PLGA and various other polymeric functionalizations could effectively improve its solubility and stability in phosphate buffered saline and physiological medium. Cs_2CO_3 and MoO_3 are used to enhance its electron transfer properties and to prevent degradation from air exposure in BP based FET biosensors. Copper functionalization via vacuum deposition as well as selenium and arsenic based mineralizer assisted gas phase transformation modification provide stability to BP nanostructures to a considerable extend. TiL_4 modified BPQDs were established to be potent competitor for photothermal therapy of invasive tumor cell ablation and an improved colloidal stability for up to 72 h was observed with considerably very low degradation rate [30]. The stability of BPQDs in physiological media was significantly improved by positively charged PEG–NH_2 modification also when compared to bare form. A rational blending of exterior surface passivation, doping and hybridisation will effectively yield air, moisture and light stable BPQDs even though available studies are very limited in this field. It is envisioned that more detailed researches in this regard would come out in the most expected nearby future itself making BPQDs a trustworthy substitute for currently employed materials.

6 Applications of Black Phosphorus Quantum Dots in Biomedicine

Black phosphorus has captured fascinating scientific interest since graphene and other 2D materials has got scientific attention among researchers worldwide. Even though black phosphorus in the form of its bulk counterpart was discovered over a century ago by Bridgeman in 1914 from white phosphorus, only since 2014 with the renaissance of graphene and related compounds, BP in its monolayer and quantum dot structure has got acceptance and remarkable status in material science and related technology. Zero dimensional nanoparticles also known as quantum dots has become an emerging class of nanomaterials possessing excellent biocompatibility and in vivo biodegradability is an outstanding candidate with stringent characteristics. Exhibiting unique photo-physical characteristics and strong photoluminescence are the typical properties of BPQDs that makes them attractive candidates in the ground of electronics and optoelectronics, photovoltaics, energy storage, sensing devices as well as biomedical technology. Compared to other carbon based nanodots and semiconductor nanocrystals, BPQDs has gained acceptance especially in the biomedical scenario because of its eccentric characteristics like excellent biocompatibility, effective biodegradability releasing phosphorus atoms and the key factor reduced cytotoxicity. Because of these fascinating uniqueness, various BP nanodimensions are been widely studied now a days for diversified biomedical applications such as biosensing, bioimaging, site specific targeted drug delivery, cancer theranostics, photothermal therapy, tissue engineering, photodynamic therapy, photoacoustic like various imaging guided therapies (PET, SPECT, MRI etc,) (Fig. 2).

All these stringent applications of BPQDs are now days the basic focus of various researchers and scientists in various acclaimed organisations all around the world. Recent literatures states that graphene and its analogues as well as transition metal dichalcogenides (TMDCs) based QDs have emerged as new suggestive for bioimaging and biosensing but researches suggests that they show low cytotoxicity only at extremely low concentrations when compared to high concentrations of BPQDs. Also lack of evidences about suggestive clearance pathways for removal of TMDCs out of the physiological system limits their wide spectral biomedical applications. Therefore BPQDs have become the best suggestive to look forward far beyond in this scenario when compared to other counterparts of 2D materials. Although the perspective of this material seems outstanding for these applications, its practical applications are still challenging with so much of complicated aspects that needs

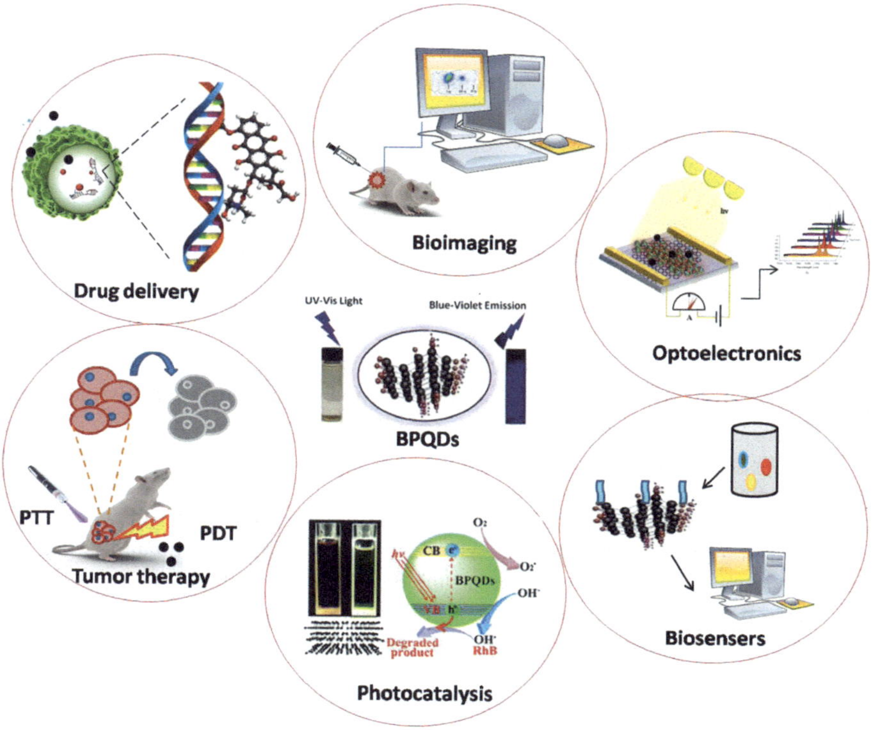

Fig. 2 Biomedical applications of black phosphorus QDs

to be unveiled. In the coming sections, all these biomedical applications will be discussed in detail one by one.

6.1 Synergistic Cancer Therapy

Regardless of sex or age group, malignant tumors have been widespread to date and nanotechnology has made considerable effort devoted to find solutions to this dreadful scenario. Various 2D materials like graphene, graphene oxide, fullerenes, transition metal dichalcogenides are been widely investigated for the same. However toxicity as well as compatibility issues makes their practical level applications complicated on an immediate effect. Comparatively a new candidate, BP has contributed an appreciable impact in this field of biomedical research, considering quantum dots and monolayer structures more efficient in this aspect. Poor stability of BP hinders its major applications, whereas for tumor therapy this instability becomes a boon which brings about a new hope in the field of biomedical technology. Interestingly after targeted action of chemotherapy and photothermal therapy, this fast degradation will leave elemental phosphorus as remnant, which is highly biocompatible and an essential element for sustenance of life. This elemental phosphorus can be further converted to phosphorus oxide in vivo; this compound is also extremely biocompatible [31]. Similar to graphene, they possess outstanding photothermal efficacy, huge surface area and size tunable narrow emission spectra that endow its craft in targeted drug delivery, photothermal therapy and imaging guided radiotherapy for combinatorial cancer theranostics [32].

Recent in-depth studies have reported that, particles having size less than 10 nm could effectively be excreted out of the system through renal routes, which suggests the applicability of QDs for biomedical applications. In this scenario, the concept of black phosphorus in quantum dot dimensionality could have been an effective strategy which can be put forward for future healthcare applications. Inorder to further enhance biocompatibility, effective surface functionalization is a sure shot criterion. For photo acoustic therapy and fluorescent imaging of live tumor cells, BPQDs have been investigated as nanotags. In a study conducted by Li et al. [33], they have fabricated PEGylated BPQDs of size less than 10 nm through liquid exfoliation top down approach, which exhibits good NIR photothermal efficiency as well as singlet oxygen generation capacity. For the first time they have incorporated both photothermal and photodynamic effect in a single material based nanoplatform for cancer treatment and successfully evaluated the performance in various cell lines and animal models. Studies revealed that, while comparing with other 2D systems, BPQDs and monolayers can hold more amount of drug moieties like doxorubicin on their surface due to enlarged surface area and other topographical features like puckered lattice structure. Being excellent photosensitizers, BPQDs can be well explored for photothermal and photodynamic therapies apart from drug delivery. BPQDs possess broad absorption spectrum and size tunable narrow fluorescence emission spectrum across the entire UV/visible region, which shows excellent NIR photothermal efficacy when compared to other similar chemical moieties. Although ultrasmall BP candidates like BPQDs show good biocompatibility and excellent in vivo biodegradability, rapid excretion and bleaching of its optical properties due to fast degradation hinders its use in clinical applications. One recent study reported an effective strategy to surpass these defects by modifying BPQDs with PLGA polymer for synergistic tumor therapy via oil in water solvent evaporation technique [34]. This integration was proved to be able to isolate BPQDs from various degradation influencing factors such as oxygen and water molecules due to hydrophobic nature of PLGA composite. This will eventually improve its photothermal efficacy and tumor ablation potency by inducing controlled degradation in the physiological system.

Photodynamic therapy is another key factor influencing tumor cell destruction usually triggered by a photosensitizer at a specific wavelength excitation. This will lead to the initiation of an energy transfer cascade that target transfer of energy to oxygen molecules in the immediate vicinity of tumor microenvironment. This will generate cytotoxic ROS which eventually kill tumor cells. Nanoparticles are nowadays scientifically accepted as good photosensitisers in this regard. The strong photoluminescent property attributed due to tunable band gap and broad absorption/narrow emission spectra, BPQDs are widely been investigated as photosensitizers for photodynamic therapy. This can therefore be used for synergistic cancer therapy. One recent study has reported that significant cell death was observed in C6 glioma and MCF-7 breast cancer cell lines treated with BPQDs along with NIR photo excitation, again clearly depicting the photothermal therapeutic potency of the material [5]. Gelation with 2D materials like graphene and graphene oxide can yield highly stable BP conjugates surpassing the major defects of BPQDs such as rapid degradation and instability to a greater extend making BPQDs suitable for various biomedical applications.

6.2 Drug/Gene Delivery

Precision and controlled delivery of drug to target site is climacteric for improved cancer therapy. Currently employed strategies are not sufficient enough to specifically address this issue. Therefore researchers are in search of novel nanotechnology mediated approaches to solve this problem. Among the reported nanomaterials, the comparatively latter addition of 2D family, black phosphorus has captivated a monumental enthusiasm among researchers worldwide to represent this scenario. One major characteristic of BPQDs, which is its high surface area, makes it admirable nominee for targeted drug/gene/nutraceutical delivery compared to currently employed strategies. Various studies have reported that bare BPQDs possess a net negative surface charge which makes it effective to incorporate positively charged drugs via electrostatic attraction. In comparison with BP nanosheets, BPQDs are a kind of more encouraging material for combinatorial therapeutic approach due to its small size and nanodot structure but very few studies have reported about this efficacy of BPQDs when compared to BP nanosheets.

More and more researches are still going on all around the world focusing on this theme. PEGylation is one basic surface modification done on BPQDs and various studies have reported about its ability to enhance physiological stability and biocompatibility. Geng et al. [35] demonstrated effective synthesis of lipid vesicle modified BPQDs for NIR mediated controlled drug release. This group formulated a BPQD-liposome conjugate in which QDs were incorporated into the hydrophobic bilayer of liposome moiety. On to the hydrophilic core of liposome moiety, therapeutic drugs were incorporated and further investigated the near infrared radiation source mediated restrained drug release profile. The encapsulated doxorubicin drug was found to be successfully released upon 808 nm laser irradiation. It was reported to be due to the photothermal conversion effect imposed by BPQDs (Fig. 3). Gene delivery is another important aspect of nanomaterial mediated theranostic applications. Recently, BPQDs based gene delivery was evaluated and demonstrated for cancer therapy. BPQDs based novel siRNA delivery was evaluated in pluripotent ovarian teratocarcinoma cells (PA 1 cells) and the study developed an outstanding potential of this semiconductor NP that shows a promising clinical application of BPQDs. It was reported that compared to commercially available gene delivery

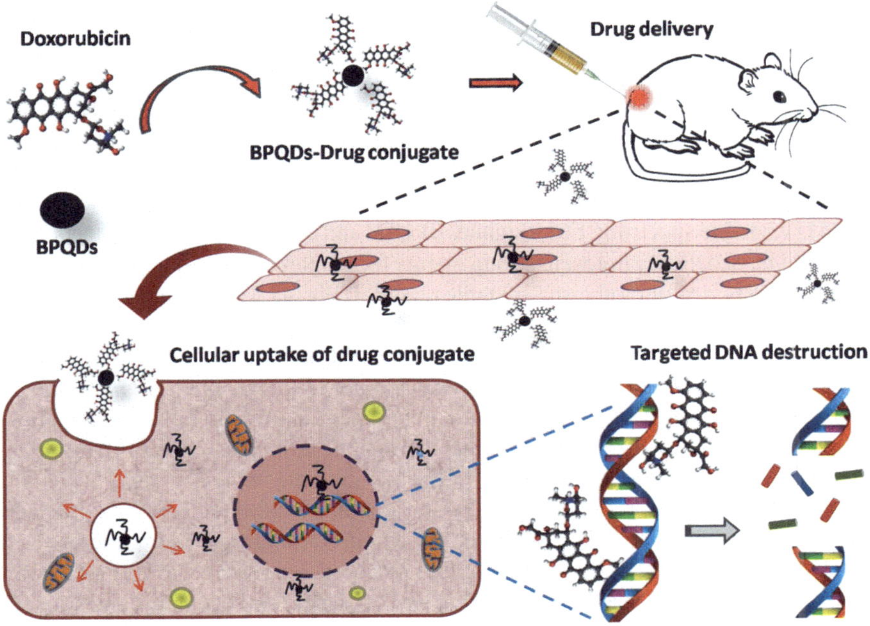

Fig. 3 Drug delivery mediated by black phosphorus QDs

agents like Oligifectamine, this BPQD composite was able to inhibit cell growth to a greater extend of about 80% [36]. Compared to conventional metallic and semiconductor QDs showing long term toxicity and bioaccumulation with heavy metal leaching, BPQDs being stable in physiological system with very low cytotoxicity and excellent biocompatibility enables its use in various biomedical applications.

6.3 Photothermal Therapy and Photodynamic Therapy

Cancer cells grow invasively, and its destruction seems to be quite laborious, especially when it comes to deeper tissues. Currently employed treatment modalities make use of ionising radiations to ensure deeper tissue penetration to kill tumor cells but at the same time this will harm the surrounding tissues once get exposed. In this context, as a substitute to time-honored conventional cancer therapy, nanotechnology provided immense contribution, especially in the milieu of photothermal (PTT) and photodynamic therapy (PDT). Photothermal therapy specifically dealt with ablation of invasive tumor cells with the aid of localised heating using NIR light absorbing material. At the same time in photodynamic therapy, excitation of a photosensitizer material, for example a quantum dot, will lead to energy transfer triggering oxygen molecules in the immediate vicinity generating reactive oxygen species, which will eventually kill tumor cells. Nanomaterials of varying dimensionalities and physico-chemical characteristics have been researched and employed for the same. Various 2D materials like graphene nanosheets, transition metal dichalcogenides etc, are gaining attention in this regard. Apart from these, the very recent member black phosphorus has provoked an equally impressive surge among researchers all around the world due to remarkable flexibility in various properties and size tunable band gap exhibiting broad absorption spectrum and narrow emission spectrum across the entire UV/visible spectrum. For instance, BPQDs and exfoliated BP nanosheets were proved to be excellent photosensitizing agents for singlet oxygen species generation as well as a reported photothermally stable material (Fig. 4). Moreover, the biodegradable property of BP into elemental forms and phosphorus oxides contributes to its biocompatibility as well. But only in the far UV and NIR region irradiation wavelength, they shows deep tissue penetration property and not in the visible light region.

A most recent research study depicted polydopamine functionalised black phosphorus quantum dots and its photothermal therapeutic ability capable for multimodal cancer theranostics. Results showed a promising outcome, which suggests its potential in tumor destruction hopefully proposed for future cancer therapy [37]. It was reported that in a study done by Sun et al. [5] approximately 2.6 nm lateral sized BPQDs were synthesised via liquid exfoliation top down approach through a combination of probe and bath sonication. It was found to be excellent candidate for

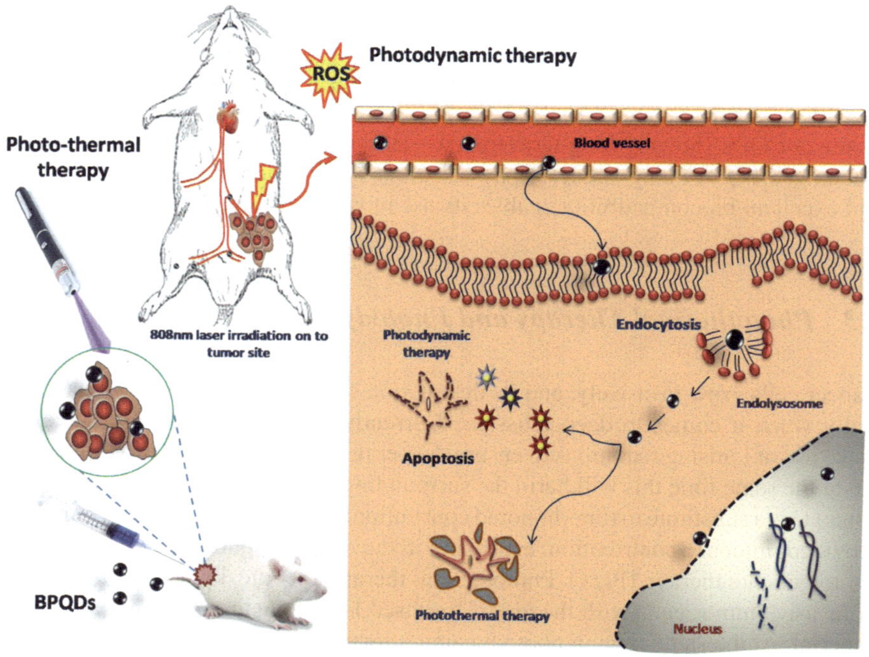

Fig. 4 Photothermal therapy and photodynamic therapy using black phosphorus QDs

NIR photothermal therapy. The PEGylated variant showed outstanding biocompatibility and extremely low cytotoxicity. In vitro studies using MCF-7 and C6 glial cells showed excellent photothermal therapeutic potency leading to significant cell death that clearly marks its ability for PTT. A multifunctional nanotheranostic platform using BPQDs was spotted by Yi Li Zhiming and coworkers in 2017 in which PEGylated BPQDs showed a combinational approach of PTT and PDT for effective tumor therapy both in vitro and in vivo. Very low cytotoxicity in cell lines and negligible level of toxicity in major internal organs was observed which provides a boost up potential for future healthcare applications [33]. Biodegradable BPQD/PLGA nanocomposite was fabricated by Shao et al. [34] which showed an enhanced photothermal efficacy and an appreciably reduced toxicity. These are evidenced as highly efficient tumor ablation agents under 808 nm laser irradiation and possess good biocompatibility as well. BPQD-biomolecule complex, BPQD-nanosheet hybrids and BPQD-polymer complexes are also under detailed investigation for various biomedical applications.

Very recently, Zhong et al. [37] reported a polydopamine functionalized BPQDs showing strong NIR absorption potency as well as excellent free radical scavenging ability that can be well attributed for PTT and PDT for a dual model tumor therapy. Ultrathin BP monolayers have been developed which shows excellent photodynamic efficacy with the release of singlet oxygen species at the same time converting itself into phosphorus oxides. Likewise, currently more and more researches are being focused on BPQD platform yet then most of the experimentations in this field are still in an outset when compared to other extensively explored 2D materials. Lack of large-scale quantum yield synthesis methods with precise control over various parameters like size, surface charge and surface modification suggest the main hindrance for BPQD researches. Therefore new suggestive methods get betrayed in the immediate future itself to counteract this strategy.

6.4 Imaging Guided Therapies

BPQDs being highly fluorescent in nature, suggests its extraordinary acceptance for imaging guided therapies like photoacoustic imaging, MRI, PET, SPECT and so on. Optical imaging therapies based on nanomaterials suggests a superlative concept for currently employed radiation-based imaging tools that impart great threat to the living system on to which it is exposed. The very recent members BPQDs and BP monolayers have shown spectacular status in this field owing to their tunable size dependent physico-chemical characteristics that makes them hold a promising status among the reported 2D materials to date. In a recent study, Sun et al. [6] reported the synthesis of PEGylated BP nanostructures from red phosphorus via a solvent less high energy mechanical milling process for photoacoustic imaging. The study well claimed that the as synthesized BP showed a considerably high yield when compared to the synthesis strategies reported till then. The in vivo photoacoustic imaging ascertained that they can effectively target and accumulate at the tumor sites through enhanced permeability retention effect. Another reported study established an NIR dependent glutathione responsive multifunctional BPQD capped Fe_3O_4@Carbon shell nanocarrier. BPQD and the carbon shell of the nanosphere shows an appreciable PTT effect while BPQD and Fe_3O_4 depicted a dual model magnetic resonance (MR) as well as fluorescence imaging. Also its empty inner core space can be loaded with drug such as doxorubicin and functionalized with an epidermal growth factor targeting aptamer for dual model imaging guided therapy [38]. Another recent study by Sun et al. [39] reported that TiL_4 functionalized BPQDs can be used as an imaging moiety for in vivo photoacoustic imaging. The study also depicted that the synthesized BPQDs were very well able to accumulate inside the tumor cells showing a high profile of spatial resolution enabling streamlined tumor visualization. More and more researches are currently focusing on QDs based imaging guided therapies such as PET, MRI and SPECT but most of all are currently under detailed investigations with lots of crucial questions arising about the potent toxic impact of the material. Researches on BPQD based imaging therapies are still in an infancy stage because it is a new member into the group of various well explored 2D materials and quantum dots. Detailed investigations and researches in this regard are undergoing in various research modalities all around the world which will hopefully get unveiled in the very nearby future itself. A very low profile of published reports exists in this context and therefore a wide gap is been opened up in this regard that attracts the focus of researchers and scientists across various parts of the world.

6.5 *Bioimaging and Biosensing*

One among the tempting area of biomedical applications of BPQDs lies in bioimaging, more precisely saying in vitro and in vivo bioimaging. BPQDs can be used as a multifunctional nanoplatform for combined imaging and imaging based therapy for malignant tumor treatment. Inorder to improve its biocompatibility furthermore, surface functionalizations can be made possible with the aid of polymeric compounds like PEG, dextran, pluronics etc, one of the fascinating properties exhibited by BPQDs include its strong fluorescence nature when excited with UV/NIR radiations. This property can be make use for bioimaging applications in biomedical technology and related nanodevices can be manufactured using this potential for detection of various bio analytes. One recent study reported the efficacy of BPQDs as a multifunctional nanoplatform for synergistic bioimaging based cancer therapy [33]. This group, for the first time reported a BPQDs based multifunctional nanoplatform using liquid exfoliation approach for bioimaging and PDT/PTT combination cancer therapy. Fluorescence imaging, photoacoustic imaging and thermal imaging are the major focussed imaging areas currently under detailed research using BPQDs. These possibilities can be used for cellular tracking, monitoring pharmacokinetics, efficient site specific drug delivery, signal transduction analysis etc, All these will eventually solve the major dilemmas current health care modalities are facing today. Various studies have reported the fluorescent based bioimaging potential of BPQDs.

BP enabled in vitro bioimaging of HeLa cells was reported in a recent study, which clearly states that the nontoxic behaviour of BPQDs in vitro makes it an excellent contrast agent for cell imaging. This can be envisioned as an encouraging strategy for drug delivery as well as cellular trafficking of target metabolites and drugs [40]. Tumor imaging is one major advantage because currently using imaging systems are widely using radiation and radioactive isotopes which always leave traces of long term side effects after therapy. Phosphorus being a vital element of our system, which constitutes the major bone mass element and the backbone component of nucleic acid makes the compound very much biocompatible. Therefore degradation will not be a curse beyond its application.

Biosensing and biosensors lays another major application aspect of BPQDs which include Field Effect Transistor biosensors (FET), Chemical gas sensors, colorimetric biosensors, electrochemical biosensors and fluorescent biosensors. Comparatively high ON-OFF current ratio of BPQDs makes then efficient candidate for FET-immunosensors for the detection of bioanalytes like antigen/antibody complexes based on electrical fluctuation and resistance. Various FET biosensors have been manufactured using different 2D materials but the specificity towards specific analyte is material dependant, therefore a comparative study based on analytes is not necessary but the efficiency of the sensor can be comparable. Due to the extreme specific characteristics of BP like high mobility, stringent in-plane anisotropy, photoconductive nature and high ON-OFF current ratio, BP has marked its potential as an excellent choice for the same. Due to easiness of effective surface functionalization and relatively high sensitivity in electrical properties, BP has marked a renaissance in the field also. Various studies have been reported and currently focussed on BP biosensors to effectively detect cancer and cardiovascular disease biomarkers. It was reported in a very recent study that BP based FET biosensors have been developed which can effectively detect immunoglobulin G with very high accuracy [41]. BP based colorimetric biosensors are on preliminary level investigations for the detection of various disease markers. Due to relatively easy and fast degradation of BP, it possesses strong electron transfer ability which can be made use for these applications. This will boost-up the precision and accuracy of colorimetry based bioassays. Fluorescent biosensors made of BPQDs also grabbed scientific attention recent times. BPQDs of blue-green fluorescence were reported to be successfully synthesised to construct a ratiometric fluorescent probe for trace mercury ion identification. The sensing system was based on the catalytic activity of mercuric ions towards Mn^{2+} ions and TPPS as well as on BPQDs [42].

Detection of carcinoembryogenic antigens from blood samples using Gold NPs conjugated BP single layer nanosheet was reported in a recent study in which BP nanosheet acts as a strong electron source for the reduction of 4-nitrophenol. This is regarded as a new suggestive for direct bench top to bed side level research strategy for clinical applications. Aptamer functionalised BP nanoplatform is recently been reported for label free detection of the cardiovascular biomarker myoglobin. Due to the enhanced electrochemical properties of the nanoconstruct and high affinity of the screened aptamer, a very high sensitivity is imparted to the investigated BP based electrochemical biosensor [43]. BPQDs are used as nanofluorophores in fluorescent biosensors. BPQDs as a fluorescent quenching material were investigated, which employs microRNA detection platform exhibiting an appreciable level of response to micro RNA [44]. One such kind was used to study acetylcholinesterase activity detection, which was demonstrated a quick and accurate probe free fluorescence sensing nanoplatform based on the mechanism of inner filter effect between 2-nitro-5-thiobenzoate (TNB) and BPQDs synthesised via sonication assisted solvothermal method [45]. Stability is the only major hindering aspect haunting researchers while considering these applications of BPQDs whereas effective passivation and surface functionalization for enhancing its stability will effectively broaden the sensing and imaging based application strategies of BPQDs.

7 Applications of BPQDs in Other Research Areas

BP emerged as a 2D nanomaterial with immense physico-chemical characteristics such as excellent thermal conductivity, electrical conductivity and optical properties. Rather than 2D layered BP nanomaterials, 0D BPQDs were introduced as a potential candidate for development of flexible memory device for the first time by Zhang et al. [4]. Since QDs are nanosized semiconductors that possess unique optical illuminations they are widely accepted in various sensing applications, bio imaging, optoelectronics and photovoltaic devices. Size dependent absorption of visible light and corresponding emission spectra of BPQDs were reported via time-dependent density functional theory calculations and it was noted that the quantum confinement effect has significant role in optical behaviour of BPQDs [1].

7.1 Sensor Applications

The unique optical property made attention to BPQDs as an alternative to other commonly used QDs such as CdSe-ZnS QDs, Graphene QDs etc., [46] for development of sensors like Electrochemical biosensors, Fluorescent biosensor etc., BPQDs synthesised via solvothermal route from bulk BP, with blue-green fluorescence under UV excitation were incorporated in a ratiometric florescence sensor for detection of mercury ions (Hg^{2+}). It was reported that the as fabricated sensor works on the basis of inner filter effect (IFE) between tetraphenylporphyrin tetrasulfonic acid (TPPS) and fluorescent BPQDs in which the fluorescence of BPQDs is quenched by TPPS. Since Hg^{2+} ions catalyses the formation of TPPS-Mn^{2+} complex, the recovery of florescence following the reaction is proportional to the concentration of Hg^{2+} ions. It was noted that the sensor works with excellent sensitivity and selectivity to Hg^{2+} ions [47]. Similarly, in a recent study by Gu et al. [42], it was reported that the IFE between green fluorescing BPQDs and 5, 5′-Dithiobis-(2-nitrobenzoic acid) can be used as a new label free sensing probe for detection of acetyl choline esterase activity via fluorescent detection of thiol groups. A sensor based on chemiluminescence is a fruitful platform for biochemical analysis. The reactive oxygen species produced react with BPQDs present in the BP QDs/H_2O_2/ClO^- system that results in the formation of excited phosphorus oxides. As a result chemiluminescence was produced at a wavelength of 530 nm [48]. $Ru(bpy)^{32+}$ is a widely used chemiluminescence system that needs various co-reactants for their high intensity light emission. A glassy carbon electrode modified with fabrication of BPQDs generated strong electrogenerated chemiluminescence (ECL) in presence of $Ru(bpy)^{32+}$. BPQDs mediated electrochemical oxidation of $Ru(bpy)^{32+}$ and associated light emission at neutral pH highlighted the future of BPQDs as a new co-reactant for $Ru(bpy)^{32+}$ dependent chemiluminescence sensor. Newly developed BPQDs incorporated chemiluminescence system shown sensitive detection of dopamine which cause decrease in ECL intensity through reacting with the oxidation product of $Ru(bpy)^{32+}$ [49].

7.2 *Optoelectronics*

Semiconducting nanomaterials have found profound applications in electronics and optoelectronics that makes huge impact in modern life. Recent additives such as MoS_2, WS_2 like TMDCs and BPQDs received much attention in the development of optoelectronic devices owing to its nonlinear saturable absorption characteristics. Unique ambipolar nature and high ON-OFF ratio of BP along with light absorption at visible as well as near infrared range make its advantage as part of optoelectronics application [50]. Recently, ultrashort laser pulse is generated in mode-locked fiber laser using BPQDs saturable absorber (BPQDs-SA). BPQDs showed significant optical saturable absorption effect when measured using aperture Z scan technique at a wavelength of 800 nm. The spectrum of soliton pulse generated from BPQDs-SA was characterized with long range stability. It was reported that BPQDs satisfy the best characteristics of an ultrafast photonic material [51]. A saturable absorber was fabricated by coating PVDF-BPQDs nanocomposite layer on the surface of aluminium foil between two fiber connectors placed in the fiber laser cavity. BPQDs with unique non-linear optical property was well studied using an Erbium Doped Soliton fiber laser. The laser showed the desired spectrum and possess excellent long-term stability, which suggested that BPQDs will be useful for ultra-short pulse propagation [52]. Likewise, an ultrafast erbium-doped fiber laser (EDF) was created using BPQDs possessing saturable absorption limit at 1550 nm. It was noted that the laser was operating in the soliton regime with formation of Kelly sidebands on the spectrum. The repeating radio frequency spectrum of the laser also indicated high stability of BPQD based mode locking. While adjusting the power of pump there generated noise-like pulse indicating potent optical illumination power of BPQDs. After adjusting the pump power to 200mW the band width was found to be raised to 24.13 nm, indicating that BPQDs can deliver a good enough optical radiance [17]. Xu [53] constructed BPQDs/polymethyl methacrylate (PMMA) composite nanofiber film via an electrospinning technique and the nonlinear optical properties were verified by femtosecond laser Z-scanning technique. Saturable absorption effect of freshly prepared and 3 months stored BPQDs/PMMA composite was found to be the same which indicate its long term stability with broad absorption range from visible to mid IR. Self-started mode locking pulses were generated after fabrication of these composite based SA in EDF. These experiments clearly depict the major space of BPQDs as an excellent candidate in the field of optoelectronics.

7.3 Photocatalysis

Photocatalysis reactions pave way to formulate antibacterial reactants, degradation of toxic organic matters for purification of waste water, purification of air etc, and are also applied in large scale production of hydrogen. Semiconducting materials with their unique electronic structure consisting of filled valance band and vacant conduction band can act as a strong catalyst in photocatalysis reaction. Semiconductors catalysts accelerate oxidation reduction reaction of organic compounds in presence of high energy light irradiation so that their degradation is getting fastened up at minimum energy expenditure [54]. Nowadays, researches are focusing on development of bandgap-tunable QDs catalysts for photocatalysis application. Recently, two BPQDs (BPQDs-NMP-Oleic acid and BPQDs-NMP) having bandgap of 2.82 and 2.96 eV were synthesised via liquid exfoliation method. They shown emission of blue and green fluorescence under UV irradiation and their photoluminescence decay curves were seemed to be fitted with excited lifetimes of 3.14 and 1.68 ns. Photocatalytic efficiency of BPQDs were analysed as measurement of total organic carbon released and temporal concentration changes of rhodamine B (RhB) under visible light. There found 92% degradation of RhB at 16 mg/L concentration of BPQDs with bandgap of 2.82 eV. Generation of highly reactive hydroxyl radical over the surface of BPQDs under visible light leads to the photocatalytic degradation of RhB through destruction of its chromophoric structure. BPQDs may be a highly efficient photocatalytic agent having novel quantum confinement and size effect [55]. Recently, a heterojunction structure of graphitic carbon nitride (g-C_3N_4) and BPQDs were synthesised in such a way that BPQDs can act as co-catalyst of g-C_3N_4 for hydrogen generation via photocatalytic water splitting.

Diffuse reflectance UV/Vis/NIR spectroscopy was performed to analyse the light absorption characteristic of bare g-C_3N_4 and BPQD- g-C_3N_4 heterostructure. With HR-TEM imaging it was noted that BPQDs layers are fabricated well over the surface of g-C_3N_4 with good interfacial charge transfer efficiency. Bare g-C_3N_4 showed a band gap of 2.7 eV whereas BPQD/g-C_3N_4 sample showed a typical bandgap of 2.58 eV which reflects in an absorption range at 480 nm. Hydrogen evolution as a result of photolytic water splitting under sunlight and LED irradiation over the surface of bare and heterostructure were compared. Electrocatalytic water splitting performance of the conjugate under light-irradiation was analysed by measuring photocurrent response generated from glassy carbon electrode coated with the same. Compared to g-C_3N_4, BPQD conjugated structure enhances the rate of hydrogen evolution. It was also observed that with increasing loading content of BPQDs there was a corresponding increase in hydrogen evolution indicating its high efficient photocatalysis property [56].

8 Toxicity of Black Phosphorus QDs

With profound development of nanotechnology research, synthetic nanomaterials attained great impact on almost all industries like medicine, energy, electronics, aerospace etc. With the prevalence of increased interest of these materials across various interdisciplinary areas, the chance of direct as well as indirect exposures of these materials to living things also increases. Even though, a bulk number of researches explored the hidden toxicity of nanomaterials, there exist many areas unexplored or under explored till now like the chronic impact of these materials to human system. With this perspective there is need for a thorough toxicity profiling of nanomaterials before their clinical and environmental applications.

High quality QDs are synthesised in large scale by industries for various biomedical and electronic applications but with very limited knowledge about their toxic impacts to biological system are available till date. It is somewhat difficult to discuss the toxicity of QDs because of the diversity in synthesis protocols, stability, physicochemical properties and mechanism of action of QDs. Size, charge, concentration, outer coating bioactivity and oxidative, photolytic, and mechanical stability also contribute a major space in toxicity determining factors of QDs. Mechanisms of QD-induced toxicity have been elucidated by studying using in vitro cell culture or by using animal models in vivo.

8.1 Toxicity Analysis Using Cell Cultures

QDs when compared with their bulk counterpart possess large surface area to volume ratio. So it is difficult to assess the toxicity of QDs with conventional assays suitable for similar semiconducting materials. Since QDs possess <10 nm size, their accessibility of interaction with cell is of great intensity. In vitro cytotoxicity studies mainly imply the interaction of nanomaterials with each and every organelle and membrane structures of cells. Recently, Lee et al. [40] investigated in vitro biocompatibility of BPQDs in three different cell lines including HeLa, COS-7, and CHO-K1. It was found that exposure of 0–3.0 mg/mL of BPQDs for 12 h doesn't cause any cytotoxic effect and was very much compatible with the biological system with little or no toxicity even at the highest selected concentration. Results were confirmed to be same with increase in exposure time to 24 and 48 h using ionized BPQDs. As it is found to be non toxic, its uptake by HeLa cells and fluorescence emission at blue-green range was also demonstrated to highlight their potential application as a bio imaging probe. Similarly, when BPQDs undergo surface stabilisation with catechol incorporated PEG become more hydrophilic and emit stable blue fluorescence. They showed meagre toxicity to KB and MDCK cell lines with excellent fluorescent live cell imaging capability [57].

Since, nanoparticles are immensely cleared from circulation by blood macrophages, the major cell responsible for inflammatory reaction associated with nanoparticle accumulation, Qu et al. [58] selected raw 264.7 macrophages cells to evaluate the fate of BPQDs and titanium sulfonate ligand (TiL_4) modified BPQDs in vitro. Raman spectroscopy analysis revealed that bare BPQDs are more likely to get internalized and distributed in cellular organelles compared to TiL_4 modified BPQDs. Cytotoxicity of both BPQDs in terms of ATP utilization and lysosomal stability was analyzed in J774A.1 and raw 264.7 cells. Survival rate of macrophages are not affected by both particles where bare BPs induced falling of cellular energy source like ATP in J774A.1 cells. Morphological evaluation of lysosomes also strengthens the finding that bare BPs caused lysosomal swelling whereas TiL_4 modified BPQDs did not induce any such morphological abnormalities. Cytokines the cellular signalling molecules that accelerate inflammatory reaction have critical role in nanoparticle mediated immune interaction. One of the major inflammatory cytokine TNF α was found to be elevated in bare BPQDs treated macrophages compared to TiL_4 modified BPQDs further confirms the better biocompatibility of TiL_4 modified BPs.

A wide range of in vitro assays such as free radical associated oxidative stress, induction of apoptosis and DNA damage were employed for assessing toxicity of BPQDs to HeLa cells by Mu et al. [7]. After synthesis of BPQDs via liquid ultrasonication exfoliation method, a detailed characterization was carried out and it was observed that the size of BPQDs lies at a range of 2.7 ± 0.7 nm. After exposure with increasing concentration of BPQDs, viability of Hela cells were reduced respectively with the dose and was attributed through oxidative stress generated inside the cells. Oxidative stress inside the cells eventually led to apoptotic or necrotic cell death. It was also cited that highly active radical molecules generated are able to react with genetic material so that it may cause DNA fragmentation in CHO cells after exposure to highest concentration (200 μg/mL) of these material.

Even if the physical nature of nanomaterial have critical role in a physiological system, it is not the major parameter that define the possible impact of those materials in a living system. It may get redefined by the formation of protein corona when the particle comes in contact with the blood stream. Protein corona formation and associated immune perturbation of BP nanosheet (BPN) and BPQDs were reported in recent times. It was observed that BPNSs and BPQDs produced slight cytotoxic effects on various cancer cell lines and macrophages from peripheral blood. Notably, when BPN and BPQDs were covered with protein corona, even if their rate of uptake by macrophages were high, their cytotoxic effect was significantly reduced. BPQD-protein corona complex showed immune modulatory effect on dTHP1 cells with induction of various cytokines like Interleukin 1beta, IL6 and interferon γ. After exposure to BP nanomaterials as well as BP–corona complexes, there observed time-dependent increase in production of reactive oxygen radicals in macrophage-like cells. Corona complex formations around the BP nanomaterials even enhance ROS production inside the cell cytoplasm. The study highlight the role of protein corona in modifying immune mechanisms and associated inflammatory illness followed by theranostic application of nanostructures [59].

Since BPQDs possess outstanding drug loading capacity and photothermal efficiency it attracts attention as a drug delivery agent along with chemotherapeutic Doxorubicin and anti inflammatory compound Kirenol fabricated in RBC membrane vehicle. In this study biocompatibility (in terms of RBC lysis) and immune clearance (as rate of Phagocytosis by RAW264.7 macrophages) of BPQDs as well as RBC-BPQDs composite were analysed prior to their in vivo photo thermal efficiency assessment. RBC-BPQDs nanocomposites showed excellent biocompatibility with $<1\%$ of haemolysis and low phagocytic clearance compared to BPQDs alone. The research concluded that membrane encapsulated BPQDs passively target the tumour cells, as an anti-inflammatory and chemotherapeutic agent [60]. Li et al. [33] adopted CCK-8 assay to determine cell viability in which the amount of formazan dye, generated by the activities of dehydrogenase enzyme in cells directly proportional to the number of living cells. Before conducting in vivo photothermal efficiency analysis, PEGylated BPQDs was treated with HepG2 cells in vitro for 24 h. It was reported that the cells maintained $>80\%$ viability even after exposure with 100 μg/ml concentration of PEGylated BPQDs. From the study it was noted that green fluorescing PEGylated BPQDs is a best candidate for cancer therapy along with bio imaging proficiency.

8.2 *Toxicity Analysis Using Animal Models*

It is very much essential to consider the safety of any nanomaterial intended for biomedical applications before their direct entry into general application. A thorough investigation of nano interaction with biological environment using specific in vivo models is necessary to recognize the potential side effects of these materials to all natural consumers including humans. One of the common in vivo models like mice were injected intravenously (I.V) with bare BPs, TiL_4 modified BPQDs or PBS (solvent control) and sacrificed at 1, 7, and 28 days to assess the potential inflammatory responses. It was noted that TiL_4 modified BPQDs treated animals exhibited no acute inflammatory response compared to bare BPQDs which accelerate intracellular cytokine production and Neutrophil activation [58]. In another study, BPQDs introduced intra peritonealy to mice and in vivo oxidative stress and apoptosis were evaluated in liver along with assessment of DNA damage in Bone marrow cells. There found transient level of DNA damage, oxidative stress and lipid peroxidation after in vivo injection of BPQDs which recovered gradually to normal levels. There found no or little inflammatory, hematological and pathological abnormalities in mice after BPQD exposure for longer duration [7].

Nowadays researchers are focussing on the formulation of nanocomposites capable of execute combined PDT and PTT which will be an effective treatment strategy for cancer in future. BPQDs having distinct optical property revolutionised the characteristics of 2D nanomaterial with effective generation of singlet oxygen.

They are designed as a theranostic agent in combination with iron oxide nanoparticles. These multi-formulations also have other applications like tumor theranostics and photo catalysis. In vivo biocompatibility of genipin-wrapped PGA-Fe_3O_4-CDs-BPQDs composite in terms of the haemolytic effect was conducted by Zhang et al. [61] and it was observed that these nanocomposites exhibit negligible haemolytic activity. To assess potential in vivo toxicity, PGA-Fe_3O_4-CDs-BPQDs nanocomposite were injected I.V to mice. In vivo fluorescence imaging of nanocomposites revealed their proper tissue distribution and associated tumor inhibition followed by laser irradiation. It was reported that PGA-Fe_3O_4-CDs-BPQDs nanocomposite was found to be safe while introducing into the in vivo system principally in terms of all haematological, immunological and physiological parameters analysed in mice (Fig. 5).

Biodegradability becomes a major problem while adopting nanobased products for medical application because of the least degradation potential of nanomaterials under natural condition. The long term existence of these particles in the body definitely accentuates the risk of deleterious effects. One study investigated tumor targeting and photo thermal efficiency of BPQDs conjugated with poly (lactic-co-glycolic acid) to accomplish high therapeutic efficacy and desirable biodegradation. After clearing in vitro cytotoxicity analysis using human skin fibroblast, MCF7 and B16 cells, systematic in vivo toxicological evaluation was carried out in mice. I.V injection of PLGA-BPQDs to mice showed no hepatic, renal or immunotoxicity. Bio distribution analysis also indicated harmless clearance of BPQDs as nontoxic phosphate and phosphonate degradation products. All these in vivo parameters can

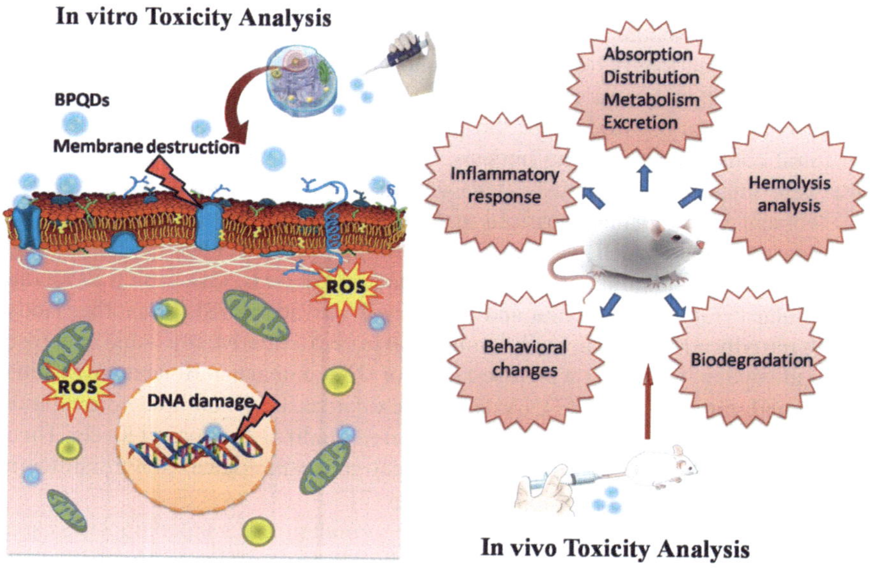

Fig. 5 Nano-biointeractions of black phosphorus QDs

potentially support BPQDs incorporated cancer treatment like PTT by specific as well as precise targeting [34].

9 Conclusion

Black phosphorus (BP) has captured fascinating scientific interest since graphene and other 2D materials has got scientific attention among researchers worldwide. Even though black phosphorus in the form of its bulk counterpart was discovered over a century ago by Bridgeman in 1914 from white phosphorus, only since 2014 with the renaissance of graphene and related compounds, BP in its monolayer and quantum dot structure has got acceptance and remarkable status in material science and related technology. It possess wide band gap than layered black phosphorus due to the renowned QD confinement effect. BPQDs have received considerable attention because of their exceptional properties applicable for vapour sensors, optoelectronics and electronic devices as well as cancer therapy. BPQDs are effective substitute for various decisive cancer theranostics due to its inherent unique characteristics. Severe instability and rapid degradation plays two major drawbacks of BP nanostructures. Effective functionalization is one major criterion to overcome the rapid degradation of this material as well as to improve its colloidal stability. Nanomaterial toxicity assessment is a crucial factor need to be assessed before validating any biomaterial for biomedical application. Very few studies are available regarding the same for BPQDs. Therefore with proper toxicity profiling, BPQDs will be a boon for next generation healthcare applications.

Acknowledgements The authors wish to express their sincere thanks to Director and Head, Biomedical Technology Wing, Sree Chitra Tirunal Institute for Medical Sciences and Technology, Trivandrum, Kerala, India for their encouragement and support for conducting this study. Anju thanks University Grants Commission, New Delhi, Prajitha thanks CSIR, New Delhi and Reshma thanks DST (Inspire fellowship), New Delhi for financial support of Junior Research Fellowships.

Conflict of Interest
The authors declare that they have no conflict of interest.

References

1. Niu, X., Li, Y., Shu, H., et al.: Anomalous size dependence of optical properties in black phosphorus quantum dots. J. Phys. Chem. Lett. **7**(3), 370–375 (2016)
2. Sofer, Z., Bousa, D., Luxa, J., et al.: Few-layer black phosphorus nanoparticles. Chem. Commun. **52**(8), 1563–1566 (2016)
3. Zhu, C., Xu, F., Zhang, L., et al.: Ultrafast preparation of black phosphorus quantum dots for efficient humidity sensing. Chem. Eur. J. **22**(22), 7357–7362 (2016)
4. Zhang, X., Xie, H., Liu, Z., et al.: Black phosphorus quantum dots. Angew. Chem. Int. Ed. **54**(12), 3653–3657 (2015)

5. Sun, Z., Xie, H., Tang, S., et al.: Ultrasmall black phosphorus quantum dots: synthesis and use as photothermal agents. Angew. Chem. Int. Ed. **54**(39), 11526–11530 (2015)
6. Sun, C., Wen, L., Zeng, J., et al.: One-pot solventless preparation of PEGylated black phosphorus nanoparticles for photoacoustic imaging and photothermal therapy of cancer. Biomaterials **91**, 81–89 (2016)
7. Mu, X., Wang, J.Y., Bai, X., Xu, F., et al.: Black phosphorus quantum dot induced oxidative stress and toxicity in living cells and mice. ACS Appl. Mater. Interfaces **9**(24), 20399–20409 (2017)
8. Bagher, A.M., Vahid, M., Mohsen, M.: Geothermal energy. J. Eng. Technol. Res. **6**, 146–150 (2014)
9. Wang, Z., Liu, Z., Su, C., et al.: Biodegradable black phosphorus-based nanomaterials in biomedicine: theranostic applications. Curr. Med. Chem. **24**, 48–70 (2017)
10. Bridgman, P.W.: Two new modifications of phosphorous. J. Am. Chem. Soc. **36**, 1344–1363 (1914)
11. Lino, M.A., de Sousa, J.S., da Costa, D.R., et al.: Charging energy spectrum of black phosphorus quantum dots. J. Phys. D Appl. Phys. **50**, 305103 (2017)
12. De Sousa, J.S., Lino, M.A., da Costa, D.R., et al.: Substrate effects on the exciton fine structure of black phosphorus quantum dots. Phys. Rev. B **96**, 035122 (2017)
13. Puzder, A., Williamson, A.J., Grossman, J.C., et al.: Computational studies of the optical emission of silicon nanocrystals. J. Am. Chem. Soc. **125**, 2786–2791 (2003)
14. Wang, X., Zhang, R.Q., Lee, S.T., et al.: Anomalous size dependence of the luminescence in reconstructed silicon nanoparticles. Appl. Phys. Lett. **93**, 243120 (2008)
15. Wang, X., Zhang, R.Q., Lee, S.T., et al.: The unusual size dependence of the optical emission gap in small hydrogenated silicon nanoparticles. Appl. Phys. Lett. **90**, 123116 (2007)
16. Lin, S., Chui, Y., Li, Y., et al.: Liquid-phase exfoliation of black phosphorus and its applications. Flat Chem. **2**, 15–37 (2017)
17. Liu, H., Sun, M., Su, Y., et al.: Chemiluminescence of black phosphorus quantum dots induced by hypochlorite and peroxide. Chem. Commun. **54**(57), 7987–7990 (2018)
18. Chen, W., Li, K., Wang, Y., et al.: Black phosphorus quantum dots for hole extraction of typical planar hybrid perovskite solar cells. J. Phys. Chem. Lett. **8**(3), 591–598 (2017)
19. Meng, M., Gan, Z.X., Zhang, J., et al.: Hierarchical self-assembly of black phosphorus quantum dots with quantum confinement effects to a centimetre scale membrane. Phys. Status Solidi B **254**, 1700011 (2017)
20. Wang, W., Niu, X., Qian, H., et al.: Surface charge transfer doping of monolayer molybdenum disulfide by black phosphorus quantum dots. Nanotechnology **27**, 505204 (2016)
21. Gui, R., Jin, H., Wang, Z., et al.: Black phosphorus quantum dots: synthesis, properties, functionalized modification and applications. Chem. Soc. Rev. **47**(17), 6795–6823 (2018)
22. Ge, S., Zhang, L., Wang, P., et al.: Intense, stable and excitation wavelength-independent photoluminescence emission in the blue-violet region from phosphorene quantum dots. Nat. Sci. Rep. **6**, 27307 (2016)
23. Wang, G., Pandey, R., Karna, S.P.: Phosphorene oxide: stability and electronic properties of a novel two-dimensional material. Nanoscale **7**(2), 524–531 (2015)
24. Anju, S., Ashtami, J., Mohanan, P.V.: Black phosphorus, a prospective graphene substitute for biomedical applications. Mater. Sci. Eng. C **97**, 978–993 (2019)
25. Zhou, Q., Chen, Q., Tong, Y., et al.: Light-induced ambient degradation of few-layer black phosphorus: mechanism and protection. Angew. Chem. Int. Ed. **55**(38), 11437–11441 (2016)
26. Ahmed, T., Balendhran, S., Karim, M.N. et al.: Degradation of black phosphorus is contingent on UV–blue light exposure. 2D Mater. Appl. **1**(1), 18 (2017)
27. Huang, Y., Qiao, J., He, K., et al.: Interaction of black phosphorus with oxygen and water. Chem. Mater. **28**(22), 8330–8339 (2016)
28. Kim, J., Baek, S.K., Kim, K.S., et al.: Long-term stability study of graphene-passivated black phosphorus under air exposure. Curr. Appl. Phys. **16**(2), 165–169 (2016)
29. Liu, Y., Zhou, M., Zhang, W., et al.: Enhanced photocatalytic properties of TiO_2 nanosheets@ 2D layered black phosphorus composite with high stability under hydro-oxygen environment. Nanoscale **11**(12), 5674–5683 (2019)

30. Zhao, Y., Wang, H., Huang, H., et al.: Surface coordination of black phosphorus for robust air and water stability. Angew. Chem. Int. Ed. **55**(16), 5003–5007 (2016)
31. Wang, H., Yang, X., Shao, W., et al.: Ultrathin black phosphorus nanosheets for efficient singlet oxygen generation. J. Am. Chem. Soc. **137**(35), 11376–11382 (2015)
32. Tao, W., Zhu, X., Yu, X., et al.: Black phosphorus nanosheets as a robust delivery platform for cancer theranostics. Adv. Mater. **29**(1), 1603276 (2017)
33. Li, Y., Liu, Z., Hou, Y., et al.: Multifunctional nanoplatform based on black phosphorus quantum dots for bioimaging and photodynamic/photothermal synergistic cancer therapy. ACS Appl. Mater. Interfaces **9**(30), 25098–25106 (2017)
34. Shao, J., Xie, H., Huang, H., et al.: Biodegradable black phosphorus-based nanospheres for in vivo photothermal cancer therapy. Nat. Commun. **7**(12967), 1 (2016)
35. Geng, S., Wu, L., Cui, H., et al.: Synthesis of lipid–black phosphorus quantum dot bilayer vesicles for near-infrared-controlled drug release. ChemComm **54**(47), 6060–6063 (2018)
36. Yin, F., Hu, K., Chen, S., et al.: Black phosphorus quantum dot based novel siRNA delivery systems in human pluripotent teratoma PA-1 cells. J. Mater. Chem. B **5**(27), 5433–5440 (2017)
37. Zhong Li, XuH, Shao, J., et al.: Polydopamine-functionalized black phosphorus quantum dots for cancer theranostics. Appl. Mater. Today **15**, 297–304 (2019)
38. Zhang, M., Wang, W., Cui, Y. et al.: Near-infrared light-mediated photodynamic/photothermal therapy nanoplatform by the assembly of Fe_3O_4 carbon dots with graphitic black phosphorus quantum dots. Int. J. Nanomed. 2803 (2018)
39. Sun, Z., Zhao, Y., Li, Z., et al.: TiL_4 coordinated black phosphorus quantum dots as an efficient contrast agent for in vivo photoacoustic imaging of cancer. Small **13**(11), 1602896 (2017)
40. Lee, H.U., Park, S.Y., Lee, S.C., et al.: Black phosphorus (BP) nanodots for potential biomedical applications. Small **12**(2), 214–219 (2016)
41. Yantao, Chen, Ren, R., et al.: Field-effect transistor biosensors with two-dimensional black phosphorus nanosheets. Biosens. Bioelectron. **89**, 505–510 (2017)
42. Gu, W., Yan, Y., Pei, X. et al.: Fluorescent black phosphorus quantum dots as label-free sensing probes for evaluation of acetylcholinesterase activity. Sens. Actuat. B: Chem. Chem. **250**(601–607) (2017)
43. Vinod, K., Brent, J.R., Shorie, M., et al.: Nanostructured aptamer-functionalized black phosphorus sensing platform for label-free detection of myoglobin, a cardiovascular disease biomarker. ACS Appl. Mater. Interfaces **8**(35), 22860–22868 (2016)
44. Zhou, J., Li, Z., Ying, M., et al.: Black phosphorus nanosheets for rapid microRNA detection. Nanoscale **10**(11), 5060–5064 (2018)
45. Wei, Y., Pei, X., Zhang, C., et al.: Fluorescent black phosphorus quantum dots as label-free sensing probes for evaluation of acetylcholinesterase activity. Sens. Actuator B-Chem. **250**, 601–607 (2017)
46. Matea, C.T., Mocan, T., Tabaran, F., et al.: Dots in imaging, drug delivery and sensor applications. Int. J. Nanomed. **12**, 5421 (2017)
47. Gu, W., Pei, X., Cheng, Y., et al.: Black phosphorus quantum dots as the ratiometric fluorescence probe for trace mercury ion detection based on inner filter effect. ACS Sensors **2**(4), 576–582 (2017)
48. Liu, M., Jiang, X.F., Yan, Y.R., et al.: Black phosphorus quantum dots for femtosecond laser photonics. Opt. Commun. **406**, 85–90 (2018)
49. Zhang, L., Tian, K., Dong, Y., et al.: Electrogenerated chemiluminescence of Ru (bpy)32+ at a black phosphorus quantum dot modified electrode and its sensing application. Analyst **143**(1), 304–310 (2018)
50. Cheng, J., Wang, C., Zou, X., et al.: Recent advances in optoelectronic devices based on 2d materials and their heterostructures. Adv. Opt. Mater. **7**(1), 1800441 (2019)
51. Xu, Y., Wang, Z., Guo, Z., et al.: Solvothermal synthesis and ultrafast photonics of black phosphorus quantum dots. Adv. Opt. Mater. **4**(8), 1223–1229 (2016)
52. Wang, Z., Xu, Y., Dhanabalan, S.C., et al.: Black phosphorus quantum dots as an efficient saturable absorber for bound soliton operation in an erbium doped fiber laser. IEEE Photon. J. **8**(5), 1–10 (2016)

53. Xu, Y., Wang, W., Ge, Y., et al.: Stabilization of black phosphorous quantum dots in PMMA nanofiber film and broadband nonlinear optics and ultrafast photonics application. Adv. Funct. Mater. **27**(32), 1702437 (2017)
54. Saravanan, R., Gracia, F., Stephen, A.: Basic principles, mechanism, and challenges of photocatalysis. In: Nanocomposites for visible light-induced photocatalysis, pp. 19–40. Springer, Cham (2017)
55. Yuan, Y.J., Yang, S., Wang, P., et al.: Bandgap-tunable black phosphorus quantum dots: visible-light-active photocatalysts. Chem. Commun. **54**(8), 960–963 (2018)
56. Kong, L., Ji, Y., Dang, Z., et al.: gC3N4 loading black phosphorus quantum dot for efficient and stable photocatalytic H2 generation under visible light. Adv. Funct. Mater. **28**(22), 1800668 (2018)
57. Lee, M., Park, Y.H., Kang, E.B., et al.: Highly efficient visible blue-emitting black phosphorus quantum dot: Mussel-inspired surface functionalization for bioapplications. ACS Omega **2**(10), 7096–7105 (2017)
58. Qu, G., Liu, W., Zhao, Y., Gao, J., et al.: Improved biocompatibility of black phosphorus nanosheets by chemical modification. Angew. Chem. **129**(46), 14680–14685 (2017)
59. Mo, J., Xie, Q., Wei, W., Zhao, J.: Revealing the immune perturbation of black phosphorus nanomaterials to macrophages by understanding the protein corona. Nat. Commun. **9**(1), 2480 (2018)
60. Huang, X., Wu, B., Li, J., Shang, Y., et al.: Anti-tumour effects of red blood cell membrane-camouflaged black phosphorous quantum dots combined with chemotherapy and anti-inflammatory therapy. Artif. Cell Nanomed. B **47**(1), 968–979 (2019)
61. Zhang, M., Wang, W., Wu, F., et al.: Black phosphorus quantum dots gated, carbon-coated Fe_3O_4 nanocapsules (BPQDs@ ssFe3O4@ C) with low premature release could enable imaging-guided cancer combination therapy. Chem. Eur. J. **24**(49), 12890–12901 (2018)

Simulation Studies for Black Phosphorus: From Theory to Experiment

Muhammad Imran, Fayyaz Hussain, Abdul Rehman, R. M. Arif Khalil, Tariq Munir, M. Zeeshan Yaqoob and Sungjun Kim

Abstract Phosphorene or 2D black phosphorus have attracted enormous attention of researcher due to its excellent structural, mechanical, electronic, magnetic and vibrational properties. This chapter presents a comprehensive review of properties of black phosphorus, techniques of improving its properties and to date most of the significant research conducted in this field of research. Studies show that most of the simulation research has been performed using molecular dynamics and density functional theory. The mechanical properties of black phosphorus have been excellent and can be tunned using defect engineering. Electronic and magnetic properties have been studied using density functional theory. It is observed that both can be successfully tuned by substituting doping of the suitable impurity atoms. Black phosphorus in its pristine form is expected to have nonmagnetic behaviour with can be revert to ferromagnetic by addition of suitable dopant which causes orbital hybridization resulting in ferromagnetism. The modes of vibrations for black phosphorene were calculated using the supercell method and these modes are characterized for IR and Raman spectroscopy. Theoretical calculated IR and Raman active modes are comparable with experimental available results.

Keywords Phosphorene · Black phosphorus · 2D materials · Ferromagnetism · DFT · Molecular dynamics

---

M. Imran (✉) · A. Rehman · T. Munir · M. Zeeshan Yaqoob
Department of Physics, Govt. College University Faisalabad, Faisalabad 38000, Pakistan
e-mail: imraniub86@gcuf.edu.pk

R. M. Arif Khalil
Materials Simulation Research Laboratory (MSRL), Department of Physics, Bahauddin Zakariya University, 60800 Multan, Pakistan

F. Hussain (✉) · S. Kim (✉)
School of Electronics Engineering, Chungbuk National University, Cheonju 28644, South Korea
e-mail: fayyazhussain248@yahoo.com

S. Kim
e-mail: lizzym19@gmail.com

Inamuddin et al. (eds.), *Black Phosphorus*, Engineering Materials,
https://doi.org/10.1007/978-3-030-29555-4_5

1 Introduction

Black phosphorus discovered about a century ago and it is also known as stable phosphorus at room temperature. It's way better stable as compared to the other allotropic forms of phosphorus [1, 2]. The use of Black phosphorus in the field of semiconductor devices came into the knowledge of researchers after the discovery of graphene which was in 2004. After that researcher who was interested in extraordinary unique properties of 2D materials in the next generations of nanodevices develop their interest in 2D materials having few layers of Black phosphorus [3, 4]. In 2013 2D Black phosphorus was extracted by mechanical exfoliation method from the bulk Black phosphorus [4].

Black phosphorus is a semiconductor having unique properties such as in bulk form its direct bandgap is about 0.3 eV and while it is considered as single layered Black phosphorus then its bandgap energy is about 2 eV. The bandgap energy and a number of layers of black phosphorus mostly remain in a linear relationship. From that relation, it can be said that bandgap can be controlled by controlling the number of layers of black phosphorus [5]. The crystal structure (CS) of Black phosphorus is orthorhombic with zigzag and armchair CS in the in-plane directions.

Van der Waals interactions are present between the layers of Black phosphorus. And the thickness of each layer of Black phosphorus is about 0.6 nm [6]. Most of the anisotropic properties of Black phosphorus is due to the orthorhombic crystal structure [7–10]. Black phosphorus has unique and physical properties which help to make Black phosphorus a best of its kind material related to the field of nanoelectronics and nanophotonic [7]. It is shown by the results obtained from molecular dynamics (MD) simulations and density functional theory (DFT) calculations that black phosphorus is tremendously anisotropic [11–15]. It's Young's modulus is quite low in the armchair direction as compared to the zigzag direction as in zigzag direction it is 56.3 and 21.9 N/m in the armchair direction [11].

The sheet of black phosphorus can resist 27% tensile strain in the zigzag direction and 30% in the armchair direction [13]. Under uniaxial tension, the temperature dependent stress-strain behavior of black phosphorus has also been studied and according to this study fracture strain, fracture strength and Young's modulus of black phosphorus steadily decreases with increasing temperature [14, 15]. As for as the defects are concern in black phosphorus there are various types of defect has been reported which strongly influence the electronic properties of black phosphorus defects which include single vacancy(SV) defects further divided into two form such as (SV1, SV2), double vacancy (DV) which is divided in three types named as (DV1, DV2 & DV3) and stone wales defects (SW) which also divided into two types such as (SW1 AND SW2) [16–18].

Mechanical response of black phosphorus is get affected by several types of defects such as Hao and Chen investigated using DFT calculations that Young's modulus can reduce in the zigzag direction by 11.7 and 6.6% of black phosphorus due to the SV2 defect and DV1 defect respectively and the armchair direction reported almost unchanged [19–21]. The strain fracture found to be reduced about 50% in the zigzag direction which is very huge reduction and this happens due to the SV2, DV1 and SW2 defects reported in a recent study of MD simulation with ReaxFF potential [20]. Sha et al. reported using MD simulations calculations with Stillinger weber potential the mechanical response of black phosphorus affected by concentration and distribution of SV2 defects [21]. Lalitha et al. investigated the relation between calcium doped phosphorene and small molecules and reported that Ca help to improve gas molecules' affinity on phosphorene [22]. Black phosphorus is taken as a most stable allotropic form of the group which include red, white and violet phosphorus [23].

2 Modern Simulation Techniques

2.1 Density Functional Theory (DFT)

DFT is a good approach to describe the ground state properties of the materials i.e. metals, semiconductors and insulator. At the microscopic level, it is used to find out the solution of many-body problems and it is known as a quantum mechanical approach. It is used to study the electronic properties of many-particle systems. It is one of the best methods used for "ab initio" calculations of the structure of the crystal. In this theory, we use functional (i.e. a function of another function) to find out the properties of many-electron systems and depend upon electron density (i.e. electron density is a basic variable). Since in this theory we use the functional of the electron density that is why this theory is known as density functional theory. It is the multipurpose computational method which is used in chemistry, computational physics and in condensed matter physics. In this theory by mapping between a non-interacting system and an interacting system of fermions we can explain many-body problems.

In materials simulation, Density functional theory is used to explain the physical and chemical behavior of many-particle systems. Here the main objective is to investigate the properties according to laws of quantum mechanics. We start with Schrödinger wave equation (SWE) and use various approximation methods to describe many-particle systems but density functional theory is still precise and effective computational method. Based on the nature of investigations Kohn-sham theory is applied distinctly. An electron gas technique is suitable for a delocalized electron in solids hence plane wave basis sets and local density type approximations are effectively used for solids. Various Exchange-correlation functionals and sophisticated functionals for chemical applications and molecular systems. Most popular functional used among researchers is Perdew–Burke–Ernzerhof (PBE) exchange model which is however inconsistent with calorimetrically accurate molecular calculations involving gas phase. The factional for chemical calculations are hybrid functional (B3LYP) in which energy from Hartree-Fock theory and exchange energy are combined.

2.2 Molecular Dynamics

MD simulation has been extensively used for modelling with atomic scale resolutions. Such atomic simulation generates 2 or 3-dimensional configurations which further need analyzation for new scientific insight. Powerful analysis software's are needed to analyze the process as the systems become much larger and complex. Molecular Dynamic Simulation was started in the early decades of 1950. In MD simulations we prepare a model of the structure, analyze the molecular properties with processes and extract some useful concept. MD simulation work on many-particle systems. MD simulation based on the force's calculation using Newtonian Physics. It consists of the solution of Newton's equation with integration and applying Tylor series to calculate the position, velocity of atoms in the system.

$$\mathrm{m_i}\ddot{r}_\mathrm{i} = \mathrm{f_i} \tag{1}$$

$$\mathrm{f_i} = -\frac{\delta U}{\delta r} \tag{2}$$

The U is the potential energy of the system that is used to calculate the interatomic forces. These interatomic potentials are calculated very precisely to get accurate results. So a proper potential is needed to calculate the interactions between particles of the system. In MD simulation Embedded atom method (EAM) potential is usually employed to calculate the metallic structures in an amorphous state. To start the simulation, every particle of the system must possess a specific position and velocity. MD simulation starts with some predefined macroscopic values, these set of thermodynamic states are known as ensembles. The various types of ensembles used commonly at different stages of the simulation process are:

1. Micro-canonical ensemble (NVE)
2. Constant volume canonical ensemble (NVT)
3. Constant pressure ensemble (NPT)

The NVE stands for the adiabatic process: no heat transfer takes place in the system. The NVE ensemble has constant volume V, Number of moles N, and constant energy E. In NVT ensemble, the temperature of the system remains constant with different types of the thermostat. Generally, Nose hoover thermostat is employed to implement temperature control during NVT and NPT ensembles.

3 Structural Analysis

Unlike graphene, Phosphorene has a 2D puckered hexagonal honeycomb structure. Usually, a vacuum surface of around 14 Å has been used along c direction to avoid unwanted interactions during the simulation studies [24]. A sample 2D black phosphorus is shown in Fig. 1.

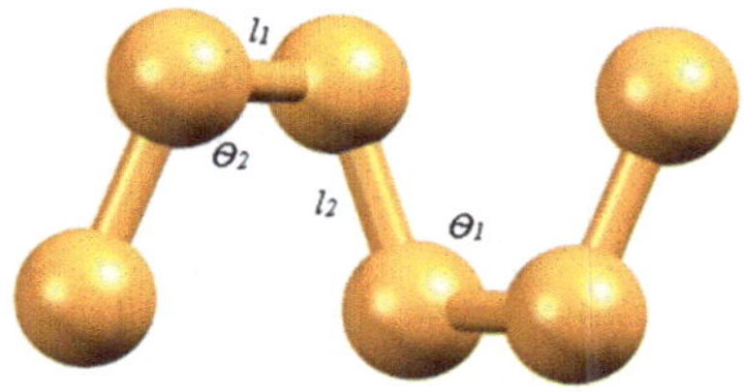

Fig. 1 The structure of 2D black phosphorus

Table 1 Optimized parameter of 2D black phosphorus [25]

Lattice parameters	a	3.314
	c	4.60
Bond angle	θ_1	103.27
	θ_2	95.63
Bond Lengths	l_1	2.230
	l_2	2.256

The optimized parameters of the 2D black phosphorus are presented in Table 1.

4 Mechanical Properties

Phosphorene has large strength and astonishing Young's modulus, hence become a potential applicant to the mechanical device. The theoretical studies predict that perfect phosphorene sheet has excellent tensile strength along armchair (3 GPa) and zigzag (6 GPa) direction. Brittle cleavage rapture and ductile fracture by plastic flow instability in molecular dynamics simulation work two specific paths of fracture in phosphorene sheet are proposed. One of the most important mechanical behavior of phosphorene shows tensile behavior in the plane so that is considered in 2D structure material because of its thickness. The intrinsic strength elastic modulus and Poisson ratio are the basic mechanical properties of phosphorene sheet which make it attractive for the researchers. In this work, we present molecular dynamics simulation of the effects of point (vacancy) defects on the mechanical behavior of phosphorene are explained based on tensile loading. The Zigzag direction has been selected to observe the effects of the tensile loading on perfect phosphorene and defected phosphorene sheet (i.e. contains mono, di, tri and tetra—vacancy). Comparison of stress versus strain data of perfect phosphorene sheet with all vacancies sheets at room temperature is also present.

To analyze the mechanical properties of pure and black phosphorus, a molecular dynamics study has been conducted. Molecular dynamics simulation code LAMMPS (Large Scale Atomic/Molecular Massively Parallel Simulator) [26] is used for all the simulations. Firstly, the perfect phosphorene sheet containing 2240 phosphorous atoms is built having a dimension of 91.952×91.8 Å along x and y-axis.

The Inter-atomic interaction has been controlled using well-known Stillinger- weber (SW) potential. The periodic boundary conditions have been employed during the simulation process in all directions. For integrating the equation of motion velocity Verlet algorithm with time step 0.5 fs is used. To minimize the energy of the system, conjugate gradient (CG) method is chosen. At prescribe, the temperature the system is equilibrated at NPT ensembles for time steps 10^5 to reduces the system stress to nearly zero. A uniaxial tensile loading with a constant rate of 0.0005 ps^{-1} is applied in along the zigzag direction to study the deformation process. These deformations take place in the armchair and Zigzag direction at 300 k at constant strain rate. The pressure is kept 0 bars in all direction to permit the free movement in Zigzag and armchair direction. The thickness of phosphorene sheet is taken about 5.24 Å. The geometric structure of undeformed and deformed perfect 2D black phosphorus is shown in Fig. 2a, b respectively.

The defected 2D Phosphorus show unique deformation behavior compared with the pristine phosphorus sheet as can be seen from Fig. 3a, b

The comparisons of stress-strain data of pristine and defected 2D phosphorus is presented below in Fig. 4.

It can be observed from the graph that pristine 2D black phosphorus shows unique and most stable mechanical configuration. The fracture stress and strain are both seems to be decreased in the presence of vacancies however the trend is not as straight forward. Sample with tetra vacancy shows least future stress among all the samples however having higher fracture strain as compared to sample with mono-vacancy. The sample with tri-vacancy shows best fracture stress compared to other defected samples. This difference in fracture stress and strain can be related to the resulting

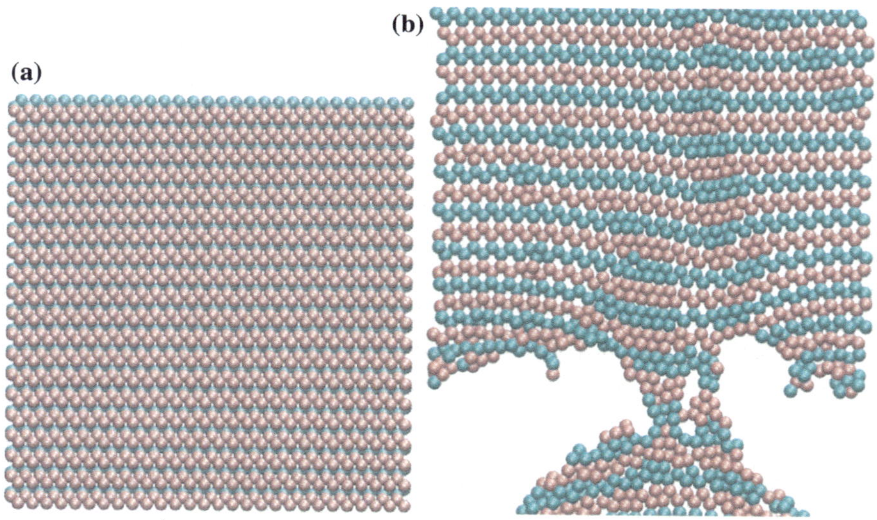

Fig. 2 The geometric structure of undeformed and deformed perfect 2D black phosphorus is shown in Fig. 2a, b respectively

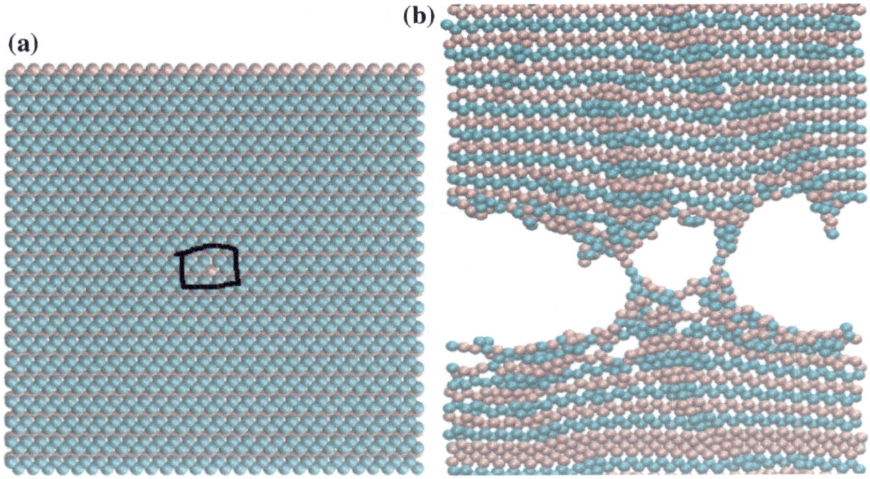

Fig. 3 The geometric structure of undeformed and deformed 2D black phosphorus containing mono-vacancy is shown in figures (**a** and **b**) respectively

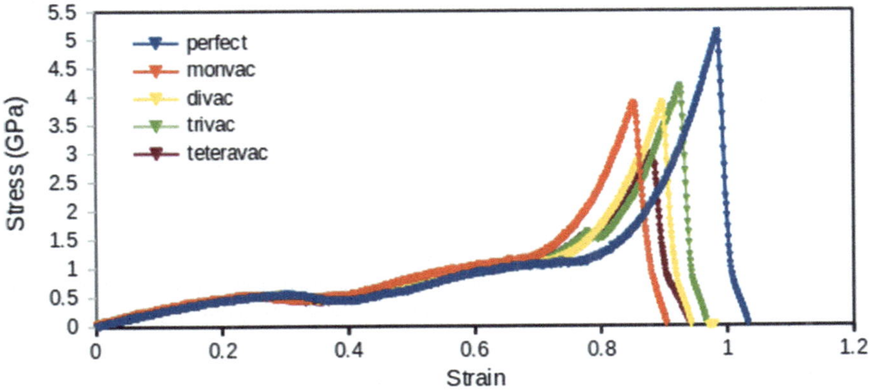

Fig. 4 The stress-strain comparison of pristine and defected 2D black phosphorus

defected configurations causes different deformation mechanism. This is the unique feature of Stone-Wales like defect formation. Similar work is also reported in 2018 [27] showing that the energy barrier for the formation of Stone-wales defects in 2D black phosphorus is much smaller as compared to graphene. The same study also predicts that the energy barrier for Stone-wales defect healing is also very high with result in a stable defected 2D black phosphorus. However, barrier height varies with various defected configuration which is responsible for the anomaly in the results presented in Fig. 4.

Molecular dynamics simulation employing the finite element method is performed to study perfect and defected phosphorene with around seven different types of defects [28]. Mechanical behaviour is being analyzed using uniaxial tensile loading technique. The results of this study reveal that the fracture stress and fracture strain can be reduced to 53 and 69% of their original values however the effect on the Young's modulus of 2D black phosphorus is quite insignificant.

A density functional tight binding study has been reported to study the failure mechanisms and deformation behavior 2D black phosphorus nanoribbons and sheets. The deformation consisting of homogenous twinning across the sheets with is responsible for a significant increase in the fracture strain of the 2D black phosphorus [29]. It is also shown that the presence of vacancy defect can cause heterogeneous nucleation with results in a reduction in the fracture strain of the twinned black phosphorus.

5 Electronic Structure Properties

In recent years just like graphene, 2D black phosphorus has also attracted enormous attention of the researchers due to their excellent and unique structural electronic, magnetic and optical properties. It has grown as a potential candidate for the nanoelectronics and optoelectronic devices. Several attempts have been reported to enhance the electronic properties of black phosphorus by the addition of suitable dopant both experimentally as well as theoretically. As the study of 2D black phosphorus is still emerging, therefore, the simulation study studies involving electronic structure properties of limited, however, the available literature is enough to enlighten the novelty of electronic structure properties of 2D black phosphorus. The tunnelable band gap of black phosphorus is layer dependent with have values varying from 0.3 to 2 eV for bulk to single layer black phosphorus [30–32]. Literature also shows that electronic band structure is anisotropic and armchair nanoribbons of black phosphorus appeared semiconductors exhibiting indirect band gap. However metallic behavior is revealed from the zigzag oriented 2D black phosphorus nanoribbons [33].

Various methods can be employed to modulate or tune the band structure of i.e. mechanical or electric straining and modification of edges etc. which makes modulating the electronic structure of 2D black phosphorus an important field of research. Some of the recently performed simulation studies involving electronic properties of black phosphorus are presented here. A first principle study has been carried out to study the electronic properties and strain induced bandgap engineering of 2D black phosphorus [25]. The simulations had been performed using a well-known density functional code quantum Espresso simulation package [34]. The Perdew–Burke–Ernzerhof (PBE) exchange model employing generalized gradient approximation (GGA) and plane augmented wave pseudopotential is used for the simulation process [35–37]. The results of band structure for in-plane uniaxial strain along zigzag and armchair direction are presented in Table 2.

It can be seen from Table 2 that results of band gap are highly anticipating for the applied strain and can shift from direct to indirect and vice versa under various

Table 2 Band structure for in-plane uniaxial strain along zigzag and armchair direction [25]

	Zigzag direction band gap (eV)	Nature of the band gap	Armchair direction band gap (eV)	Nature of the band gap
−13	0	Indirect	–	–
−9	0.3	Indirect	0	Indirect
−5	0.6	Indirect	0.6	Indirect
0	0.9	Direct	0.9	Direct
5	1.1	Direct	1.1	Direct
9	0.8	Indirect	1.2	Indirect

% strain values. The indirect band gap of zigzag and armchair 2D black phosphorus even becomes zero at −13 and −9% strain respectively.

To study the effect of doping on the electronic properties of black phosphorus nanoribbons a first principle study has been carried out in 2017 [30]. This work has been performed using Vienna Ab initio simulation package (VASP) [38] employing PAW scheme and GGA based PBE functional [35]. According to the number of valance electrons two groups of dopants has been selected to be doped along the edges and center.

1. With an odd number of valance electrons i.e. B, N, F and Al
2. Even number of valance electrons i.e. C, O, Si and S

It was found that the bond between centre doped atom and phosphorene atom is more stable as compared to edge doped atoms. The doping of atoms with an odd number of valance electrons hardly have any significant effect of the electronic states of conduction and valance band and still, nanoribbons behave as a nonmagnetic semiconductor. However, the addition of dopants with even number of valance electrons strongly affects the electronic states especially when doped near the edge of nanoribbon and show a significant and stable magnetic moment of the order of 1 $\mu_{B.}$ It is contributed to several orbits between P atoms and dopant added at the edges.

Heterostructures of 2D black phosphorus with graphene and hBN nanosheet are also gaining significant attention due to their excellent electronic properties. A DFT based simulation study has been conducted on the electronic properties of graphene and the hBN substrate supported black phosphorus. This work has also been performed using Vienna Ab initio simulation package (VASP) [39] employing PAW scheme and GGA based PBE functional. The results describing interlayer distance and bandgap relationship are presented in Table 3.

It can be clearly depicted from Table 2 that the hBN supported black phosphorus exhibit semiconducting nature with a band gap of the range 0.91 to 1.23 eV. However, the graphene supported phosphorene show metallic nature.

Table 3 Distance versus band gap relationship for graphene and hBN supported black phosphorus [40]

Interlayer distance (Å)	The band gap of hBN supported (eV)	Interlayer distance (Å)	The band gap of hBN supported (eV)
2.37	0.91	2.57	0 (metallic)
2.84	1.23	3.51	0 (metallic)
3.08	1.19	3.78	0 (metallic)
4.01	0.95	–	–

6 Magnetic Properties

Due to their excellent electronic, mechanical optical and vibrational properties researcher have an enormous interest in exploring the magnetic properties of 2D black phosphorus. Like most of the 2D materials such as graphene, silicene and MoS_2 phosphorene does not exhibit magnetism in its pristine form. A lot of effort has been made to tune the magnetic properties of black phosphorous both experimentally and theoretically. Most of the studies have been based on inducing magnetism by the addition of suitable dopant in the pristine 2D black phosphorus. Some of the most recent studies conducted for this purpose using simulation method are presented here.

To study the effect of doping on the electronic properties of black phosphorus nanoribbons a first principle study has been carried out in 2017 [30]. The addition of dopants with even number of valance electrons (i.e. C, O, Si and S) strongly affects the electronic states especially when doped near the edge of nanoribbon and show a significant and stable magnetic moment of the order of 1 μ_B. It is contributed to several orbits between P atoms and dopant added at the edges. Recently we have studied the magnetic properties of 2D black phosphorus by doping the Ce and Ti as a dopant impurity. The work has been performed using density functional theory employing VASP simulation package. Employing PAW scheme and GGA based PBE functional [24]. The doping scheme can be understood in Fig. 5.

The results of doping can be analyzed from the density of state versus energy plot as shown in Fig. 6.

The results reveal that subsequent ferromagnetism can is achieve by the doping of Ce and Ti dopant with various combinations. The calculated results depict that the pure black phosphorus shows no magnetism whereas Ti, Ce, Ti–Ti, Ce–Ce and Ce–Ti doping configurations show 0.989, 0.001, 1.100, 1.300 and 0.009 respectively. Which show strong ferromagnetism in case of Ti, Ti–Ti and Ce–Ce dopant.

The magnetic properties of black phosphorus have been investigated by substituting 4d series (Y, Zr, Nb, Mo, Tc, Ru, Rh, Pd, Ag and Cd) transition metals. The study employing first principle calculations using VASP simulation package employing PBE functional and GGA approximation. The results of the study revealed that Zr, Mo, Ru, Pd and Ag substitution induces magnetic moment μ_B in the black phosphorus as 0.148, 0.093, 0.114, 0.301 and 0.391 respectively [40].

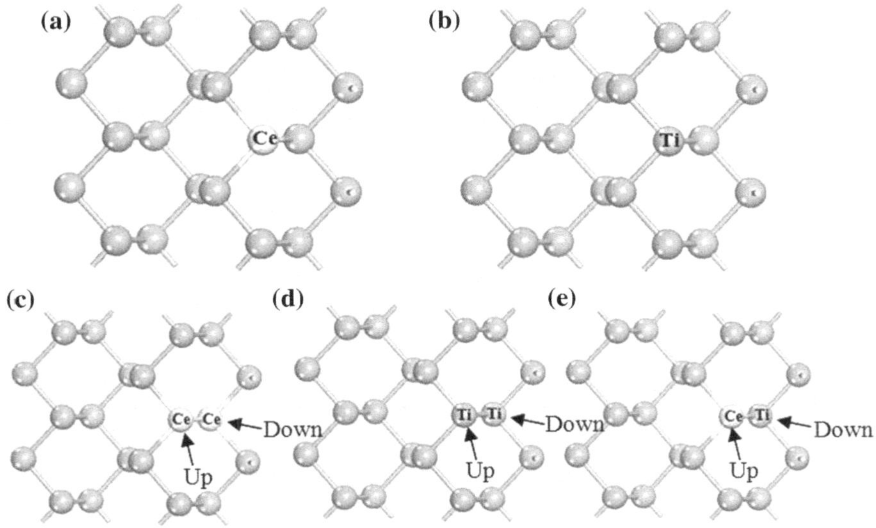

Fig. 5 **a–e** Representing Ce and Ti doping at various possible sites

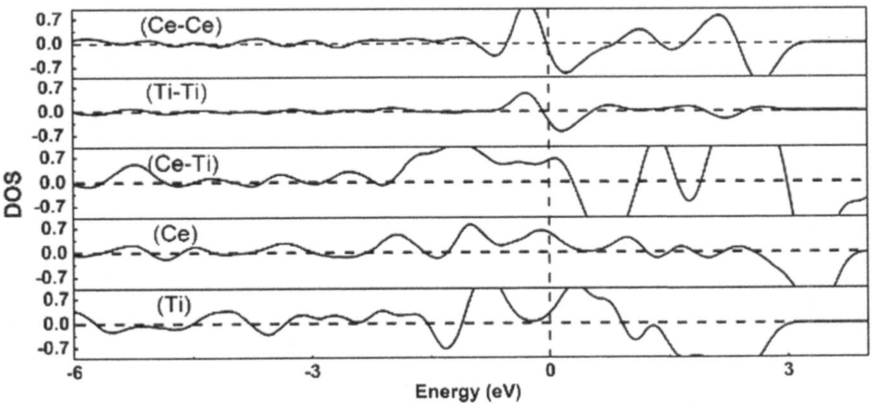

Fig. 6 The spin-polarized DOS of doped phosphorene using GGA + U functionals. Where Fermi energy (E_F) set at E = 0 eV

7 Vibrational Properties

The vibrational and thermodynamic response of phosphorous may lead to the thermal conductivity. Further structural and thermal stability of different phases of phosphorous can be analysed using the free energy calculations. The finite displacement supercell method [41] using the CASTEP code [42] adopted to find the frequencies for a supercell with dimension of 5 × 3 × 1 in case of black Phosphorene.

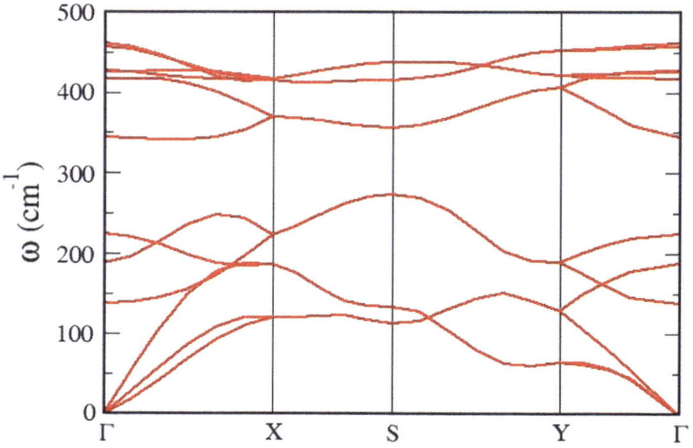

Fig. 7 Dispersion curve of black phosphorene

Table 4 Phonon modes at Γ (centre of Brillouin zone) for black Phosphorene

Irreducible representation	Character of modes	Theoretical frequency (cm^{-1}) [43]	Experiment (cm^{-1}) [44, 45]
A_g	Raman	344.1	362, 362.5
B_{1g}	Raman	427.8	439, 439.8, 439.3
A_g	Raman	457	467, 467.1, 467.6

The calculated dispersion curve of black phosphorene [43] shown in Fig. 7 is free from imaginary frequencies which show the structural stability of phosphorene structure. Table 4 shows the comparison of frequencies obtained by ab initio methods and experimental results. The shear modes of lattice vibrations are mostly observed in bulk phosphorous. However, bending modes are found in single layer phosphorene. The highest optical out of phase stretching mode in black Phosphorene are observed at frequency 461 cm^{-1}. The bending modes are found at frequency 344 cm^{-1}. The low frequency out of phase optical modes at frequency 225.2 and 189 cm^{-1} are reported but mode at 137.6 cm^{-1} is in phase in nature.

8 Conclusion

This chapter presents the simulation studies of 2D black phosphorus. The Molecular dynamics and density functional theory are the effective tools for the study of structural, mechanical, electrical, magnetic and vibrational properties of 2D black phosphorus. It can be revealed from the mechanical properties that pristine phosphorene shows good fracture strength and exhibits brittle fracture. Further analysis shows that the presence of vacancy stone-wales defects reduces the fracture stress and strain however the effect on Young's modulus is less significant. Existence of higher values of the energy barrier for the defects shows that defected black phosphorus is stable. The electronic properties of black phosphorus are orientation dependent. The armchair nanoribbons of black phosphorus appeared semiconductors exhibiting indirect band gap. However metallic behavior is revealed from the zigzag oriented 2D black phosphorus nanoribbons. Electronic properties can be tuned using strain. The black phosphorus appeared not magnetic in its pristine form however magnetism can be induced by the addition of suitable dopant. The presence of suitable dopant such as 4d series can effectively induce ferromagnetism due to orbital hybridization. Theoretical calculated IR and Raman active modes are comparable with experimental available results.

References

1. Bridgman, P.: Two new modifications of phosphorus. J. Am. Chem. Soc. **36**, 1344–1363 (1914)
2. Keyes, R.W.: The electrical properties of black phosphorus. Phys. Rev. **92**, 580 (1953)
3. Eda, G., Fanchini, G., Chhowalla, M.: Large-area ultrathin films of reduced graphene oxide as a transparent and flexible electronic material. Nat. Nanotechnol. **3**, 270 (2008)
4. Novoselov, K.S., Geim, A.K., Morozov, S.V., Jiang, D., Zhang, Y., Dubonos, S.V., Grigorieva, I.V., Firsov, A.A.: Electric field effect in atomically thin carbon films. Science **306**, 666–669 (2004)
5. Tran, V., Soklaski, R., Liang, Y., Yang, L.: Layer-controlled band gap and anisotropic excitons in few-layer black phosphorus. Phys. Rev. B **89**, 235319 (2014)
6. Zhang, R., Zhang, Y., Yu, H., Zhang, H., Yang, R., Yang, B., Liu, Z., Wang, J.: Broadband black phosphorus optical modulator in the spectral range from visible to mid-infrared. Adv. Optic. Mater. **3**, 1787–1792 (2015)
7. Xia, F., Wang, H., Jia, Y.: Rediscovering black phosphorus as an anisotropic layered material for optoelectronics and electronics. Nat. Commun. **5**, 4458 (2014)
8. Qiao, J., Kong, X., Hu, Z.-X., Yang, F., Ji, W.: High-mobility transport anisotropy and linear dichroism in few-layer black phosphorus. Nat. Commun. **5**, 4475 (2014)
9. Fei, R., Yang, L.: Strain-engineering the anisotropic electrical conductance of few-layer black phosphorus. Nano Lett. **14**, 2884–2889 (2014)
10. Luo, Z., Maassen, J., Deng, Y., Du, Y., Garrelts, R.P., Lundstrom, M.S., Peide, D.Y., Xu, X.: Anisotropic in-plane thermal conductivity observed in few-layer black phosphorus. Nat. Commun. **6**, 8572 (2015)
11. Jiang, J.-W., Park, H.S.: Mechanical properties of single-layer black phosphorus. J. Phys. D Appl. Phys. **47**, 385304 (2014)
12. Hu, T., Han, Y., Dong, J.: Mechanical and electronic properties of monolayer and bilayer phosphorene under uniaxial and isotropic strains. Nanotechnology **25**, 455703 (2014)
13. Wei, Q., Peng, X.: Superior mechanical flexibility of phosphorene and few-layer black phosphorus. Appl. Phys. Lett. **104**, 251915 (2014)
14. Sha, Z.-D., Pei, Q.-X., Ding, Z., Jiang, J.-W., Zhang, Y.-W.: Mechanical properties and fracture behavior of single-layer phosphorene at finite temperatures. J. Phys. D Appl. Phys. **48**, 395303 (2015)
15. Yang, Z., Zhao, J., Wei, N.: Temperature-dependent mechanical properties of monolayer black phosphorus by molecular dynamics simulations. Appl. Phys. Lett. **107**, 023107 (2015)
16. Hu, W., Yang, J.: Defects in phosphorene. J. Phys. Chem. C **119**, 20474–20480 (2015)

17. Farooq, M.U., Hashmi, A., Hong, J.: Anisotropic bias dependent transport property of defective phosphorene layer. Sci. Rep. **5**, 12482 (2015)
18. Li, X.-B., Guo, P., Cao, T.-F., Liu, H., Lau, W.-M., Liu, L.-M.: Structures, stabilities, and electronic properties of defects in monolayer black phosphorus. Sci. Rep. **5**, 10848 (2015)
19. Hao, F., Chen, X.: First-principles study of the defected phosphorene under tensile strain. J. Appl. Phys. **120**, 165104 (2016)
20. Xiao, H., Shi, X., Hao, F., Liao, X., Zhang, Y., Chen, X.: Development of a transferable reactive force field of P/H systems: application to the chemical and mechanical properties of phosphorene. J. Phys. Chem. A **121**, 6135–6149 (2017)
21. Sha, Z.-D., Pei, Q.-X., Zhang, Y.-Y., Zhang, Y.-W.: Atomic vacancies significantly degrade the mechanical properties of phosphorene. Nanotechnology **27**, 315704 (2016)
22. Lalitha, M., Nataraj, Y., Lakshmipathi, S.: Calcium decorated and doped phosphorene for gas adsorption. Appl. Surf. Sci. **377**, 311–323 (2016)
23. Li, L., Yu, Y., Ye, G.J., Ge, Q., Ou, X., Wu, H., Feng, D., Chen, X.H., Zhang, Y.: Black phosphorus field-effect transistors. Nat. Nanotechnol. **9**, 372 (2014)
24. Hussain, F., Imran, M., Rana, A.M., Ismail, M., Arif Khalil, R.M., Sattar, M.A., Javid, M.A., Majid, A., Cai, Y.: Phys. E: Low-dimension. Syst. Nanostruct. **106**, 352–356 (2019)
25. Phuc, H.V., Hieu, N.N., Ilyasov, V.V., Phuong, L.T.T., Nguyen, C.V.: First principles study of the electronic properties and band gap modulation of two-dimensional phosphorene monolayer: effect of strain engineering. Superlattices Microstruct. **118**, 289–297 (2018)
26. Plimpton, S.J.: Fast parallel algorithms for short-range molecular dynamics. J. Comput. Phys. **117**, 1–19 (1995)
27. Rezaee, A.E., Kashi, M.A., Baktash, A.: Stone-wales like defects formation, stability and reactivity in black phosphorene. Mater. Sci. Eng., B **236–237**, 208–216 (2018)
28. Nguyen, D.T., Le, M.Q., Nguyen, V.T., Bui, T.L.: Effects of various defects on the mechanical properties of black phosphorene. Superlattices Microstruct. **112**, 186–199 (2017)
29. Sorkin, V., Cai, Y.Q., Srolovitz, D.J., Zhang, Y.W.: Mechanical twinning in phosphorene. Extreme Mech. Lett. **19**, 15–19 (2018)
30. Zhou, W., Zou, H., Xiong, X., Zhou, Y., Liu, R., Ouyang, F.: Doping effects on the electronic properties of armchair phosphorene nanoribbons: a first-principles study. Phys. E **94**, 53–58 (2017)
31. Tran, V., Soklaski, R., Liang, Y., Yang, L.: Layer-controlled band gap and anisotropic excitons in few-layer black phosphorus. Phys. Rev. B **89**, 235319 (2014)
32. Çakır, D., Sahin, H., Peeters, F.M.: Tuning of the electronic and optical properties of single-layer black phosphorus by strain. Phys. Rev. B **90**, 205421 (2014)
33. Carvalho, A., Rodin, A.S., Neto, A.H.C.: Phosphorene nanoribbons. EPL-Europhys. Lett. **108**, 47005 (2014)
34. Giannozzi, P., Baroni, S., Bonini, N., Calandra, M., Car, R., Cavazzoni, C., Ceresoli, D., Chiarotti, G.L., Cococcioni, M., Dabo, I., et al.: Quantum espresso: a modular and open-source software project for quantum simulations of materials. J. Phys.: Condens. Matter **21**(39), 395502 (2009)
35. Perdew, J.P., Burke, K., Ernzerhof, M.: Generalized gradient approximation made simple. Phys. Rev. Lett. **77**(18), 3865 (1996)
36. Perdew, J.P., Chevary, J.A., Vosko, S.H., Jackson, K.A., Pederson, M.R., Singh, D.J., Fiolhais, C.: Atoms, molecules, solids, and surfaces: applications of the generalized gradient approximation for exchange and correlation. Phys. Rev. B **46**(11), 6671 (1992)
37. Ernzerhof, M., Scuseria, G.E.: Assessment of the perdewe burkee ernzerhof exchange correlation functional. J. Chem. Phys. **110**(11), 5029e5036 (1999)
38. Kresse, G., Furthmüller, J.: Efficient iterative schemes for ab initio total-energy calculations using a plane-wave basis set. Phys. Rev. B **54**, 11169 (1996)
39. Davletshin, A.R., Ustiuzhanina, S.V., Kistanov, A.A., Saadatmand, D., Dmitriev, S.V., Zhou, K., Korznikova, E.A.: Electronic structure of graphene– and BN–supported phosphorene. Phys. B: Condens. Matter **534**, 63–67 (2018)

40. Wanga, Y., Songb, N., Donga, N., Zheng, Y., Yanga, X., Jianga, W., Wang, B.X.J.: Electronic, magnetic properties of 4d series transition metal substituted black phosphorene: a first-principles study. Appl. Surf. Sci. **480**, 802–809 (2019)
41. Alfè, D.: A program to calculate phonons using the small displacement method. Comput. Phys. Commun. **180**, 2622–2633 (2009)
42. Clark, S.J., Segall, M.D., Pickard, C.J., Hasnip, P.J., Probert, M.J., Refson, K., Payne, M.C.: First principles methods using CASTEP. Z. Kristallogr. **220**, 567 (2005)
43. Khalil, R.M.A., Ahmad, J., Rana, A.M., Bukhari, S.H., Jamil, M.T., Tehreem, T., Nissar, U.: First principles investigation of structural, vibrational and thermal properties of black and blue phosphorene. Int. J. Mod. Phys. B **32**, 1850151 (2018)
44. Brent, J.R., Savjani, N., Lewis, E.A., Haigh, S.J., Lewis, D.J., O'Brien, P.: Production of few-layer phosphorene by liquid exfoliation of black phosphorus. Chem. Commun. **50**, 1338 (2014)
45. Guo, Z., Zhang, H., Lu, S., Wang, Z., Tang, S., Shao, J., Sun, Z., Xie, H., Wang, H., Yu, X.F., Chu, P.K.: From black phosphorus to phosphorene: basic solvent exfoliation, evolution of raman scattering, and applications to ultrafast photonics. Adv. Funct. Mat. **25**, 6996 (2015)

Biomedical Applications of Black Phosphorus

Sashivinay Kumar Gaddam, Ramyakrishna Pothu, Aditya Saran and Rajender Boddula

Abstract Two dimensional (2D) materials have attracted extravagant eminence in the area of biomedical applications due to their distinctive structure, biocompatibility, and physicochemical properties. Mono-layered black phosphorus (BP), also known as phosphorene, is a thermodynamically stable allotrope of phosphorous and is the very recent member of the 2D family has lured tremendous scientific interest since its rediscovery in 2014. The exceptional properties of BP including high carrier mobility, large specific surface area, tunable band gap, intrinsic anisotropy, and inherent in vivo biocompatibility and biodegradability make it an ideal alternative to other 2D materials in biomedical applications. This chapter summarizes various biomedical applications of BP including drug delivery, bio-imaging, biosensing, photothermal/photodynamic therapy, and theranostics. The main focus of this chapter is to emphasize the efficacy of BP nanosheets (NSs) and quantum dots (QDs) as robust and tunable diagnostic and therapeutic platforms. Finally, the current challenges and future perspectives in biomedical applications of BP have also been addressed.

Keywords Black phosphorous · Biomedical · Theranostics · Drug delivery · Bio-sensing · Imaging

---

S. K. Gaddam
LAMTUF Plastics Private Ltd, Hyderabad, Telangana, India

R. Pothu
College of Chemistry and Chemical Engineering, Hunan University, Changsha 410082, People's Republic of China

A. Saran
Department of Microbiology, Marwadi University, Rajkot, India

R. Boddula (✉)
CAS Key Laboratory of Nanosystem and Hierarchical Fabrication, National Center for Nanoscience and Technology, Beijing 100190, People's Republic of China
e-mail: research.raaj@gmail.com

Inamuddin et al. (eds.), *Black Phosphorus*, Engineering Materials,
https://doi.org/10.1007/978-3-030-29555-4_6

1 Introduction

Despite the great advancement in the field of biomedicine and therapeutics, still, human beings are under threat from various kinds of health problems like cancer, diabetes, arthritis, heart disorder, and asthma. The stressful lifestyle and reckless food habits adopted by many people are the major contributing factors to these chronic diseases [1]. Furthermore, environmental changes and ecological disturbances, due to global warming, greenhouse effect, and deforestation have been accompanied in worldwide increase in morbidity and death from a number of bloodsucking diseases like dengue, malaria, chikungunya, zika, Ebola, among others [2]. Thus, many investigations have been focused on the design and development of new smart materials for biomedical applications including pathogen detection, drug delivery, and medical imaging. Nanotechnology is a multidisciplinary field of research that deals with a variety of nanomaterials and merging nanotechnology with medicine offers a promising way to overcome the existing limitations in biomedicine [3].

Of all the nanomaterials known, carbon-based materials such as zero-dimensional (0D) fullerenes, one dimensional (1D) carbon nanotubes (CNTs), and two dimensional (2D) graphene sheets have been investigated intensively as drug carriers in modern health care applications owing to their eccentric biological and physical properties [4–11]. However, after the successful isolation of one atomic thick graphene layer via mechanical stripping in 2004, researchers have recognized the significance of "dimensionality" of a nanomaterial and it was sparked an explosion of interest in the search for additional 2D materials. This search has led to invention of different types of 2D materials in the past few years including hexagonal boron nitride (h-BN), layered double hydroxides (LDHs), covalent-organic frameworks (COFs), metal-organic frameworks (MOFs), transition metal dichalcogenides (TMDCs), transition metal oxides (TMOs), and block phosphorous (BP) [12].

2D materials comprise a high surface to volume ratio as nicely as anisotropic physical and chemical properties and as a result of these; they have pulled in an incredible scientific interest in the domain of biomedicine. Due to their high specific surface area, 2D materials can adsorb a large number of drug molecules and facilitate better control over release kinetics which makes them invaluable for drug delivery applications. Additionally, the high modulus of these materials makes them potential for improving the mechanical strength of the biomedical implants. Furthermore, these are the thinnest materials known, and hence, rapidly respond for external signals such as light which led them as potential candidates for all kinds of optical therapies, including phototherapy, sensing, and imaging. However, the practical applications of 2D materials in drug delivery system have been restricted so far due to their serious limitations. For example, even having high carrier mobility, graphene applications are limited due to its zero band gap. In contrast, TMDCs with a finite band gap have a limitation of low carrier mobility. Applications of both h-BN and LDHs have been confined due to their limited visible region absorption [13]. Hence, there is no 2D material is fully effective and were failed to meet a required comprehensive performance which is significantly important for modern therapeutic strategies.

Fortunately, the rediscovery of atomically thin BP (phosphorene) with a unique puckered single layered geometry has created a new promising alternative to these existing 2D materials for clinical applications. The novel properties of BP, such as (i) tunable bandgap from 0.3 eV (bulk) to 2.0 eV (single layer) [14–17], (ii) strong absorption in the ultraviolet (UV) and near-infrared (NIR) regions [18–21], (iii) inherent biocompatibility and great biodegradability in the physiological environment [22, 23]; that are distinguished it from other 2D materials for biomedical applications. Due to its unique properties mentioned above, BP has been drawn tremendous attention for a variety of clinical applications in biomedicine, including phototherapy, drug delivery, bioimaging, biosensing, and theranostics [24].

A great number of research articles and very few reviews have been published on biomedical applications of BP over time, but scarcely books are available to provide the information in this booming area as yet. The applications of BP in biomedicine are literally exploding, with just minimal opportunity to catch all the current activities of the field in a balanced fashion. This chapter merged the biomedical applications of BP in two aspects: (i) the unique and exceptional characteristics; (ii) the applications of BP in cancer therapy, diagnosis, and theranostics. The existing challenges and future perspectives of BP are briefly reviewed at the end of the chapter.

2 Properties of BP for Biomedical Applications

2.1 Carrier Mobility and ON-OFF Current Ratio

BP exhibits good carrier mobility (1000 $cm^2V^{-1}s^{-1}$ for a low thickened (10 nm) BPNS) that is higher than TMDCs (100 $cm^2V^{-1}s^{-1}$ for MoS_2). This property could make BP extremely sensitive to electrical perturbation and detect the adsorption as well as desorption of lethal gasses. Thus BP can be utilized in gas sensors (based upon electrical conductivity measurement) and immunosensors (with the help of electrical resistance quantification). Furthermore, BP exhibits a high ON-OFF current ratio which has resulted in applications in field-effect transistor (FET) sensing.

2.2 Tunable Band Gap

A tunable and layer dependent band gap of BP (~0.3 eV for bulk BP, ~1.88 eV for bilayer BP, and ~2.0 eV for monolayer BP) contributes to its broad energy absorption across the UV, visible, and IR regions [15]. Due to this wide tuning range, BP exhibits superior optical properties than other 2D nanomaterials, which helps in colorimetric and fluorescent analysis of different biomolecules (e.g., nucleic acids, proteins, and inorganic ions). Owing to this unique feature, BP plays a vital role in phototherapy, biosensing, bioimaging, drug delivery, etc. [16].

2.3 Anisotropy

The high in-plane anisotropic properties of BP can make it more potential over other 2D materials. Mechanical properties are one among the major anisotropic properties of BP and can bestow effectively to its clinical applications. The orthorhombic crystalline structure of BP has puckered arrangement along armchair direction but seems like a bilayer geometry across the zigzag direction. BP can endure tensile strains of up to 30% and 27% along the armchair and zigzag directions respectively [25]. Furthermore, BP has a Young's modulus of about 44 and 166 GPa along the armchair and zigzag directions respectively [26]. Because of its direction dependent Young's modulus which is one order of magnitude smaller than graphene (1 TPa) [27] or MoS_2 (270 GPa), BP exhibits outstanding mechanical flexibility over other 2D materials which makes it an attractive material for engineering strain. Moreover, BP can be utilized for a variety of biosensing applications due to its astonishing anisotropic electrical conductance and electron mobility [28].

2.4 Biocompatibility

The cytotoxicity or biocompatibility of 2D materials depends on material type, size, and concentration. Large BP (~884 $\pm$ 102.2 nm) exhibits lower biocompatibility than small BP (~208.5 $\pm$ 46.9 nm) [29]. Generally, BP possesses less cytotoxicity compared to the graphene but exhibits higher cytotoxicity than MoS_2 and WSe_2 [30]. It was found that BP also persuades cytotoxicity when it was employed at an excessive concentration (>50 μg/mL). Overall, BP displays low cytotoxicity and hence, it is suitable for clinical applications.

2.5 In Vivo Biodegradability

It is necessary to optimize the biodegradability of a 2D material before being used for biomedical applications. Generally, other 2D materials readily accumulate in physiological environment, which results in severe cytotoxicity. Hence, those materials require additional functionalization with biocompatible materials such as PPG, chitosan, glutathione, etc. to enable effective body clearance. But, experimental studies revealed that BP has high reactivity with oxygen and water [16] and can degrade in aqueous media inside the human body. Moreover, the final degradation products are non-toxic and biocompatible intermediates, like phosphates, and phosphonates which are well tolerated by the human body [22]. This inherent biodegradability makes BP as an ideal nanomaterial for in vivo clinical experiments.

In brief, the tunable band gap, superior carrier mobility, high ON-OFF current ratio, inherent biocompatibility, and in vivo biodegradability of BP suggesting its eminent potential in various types of biomedical applications such as phototherapy, drug delivery, tissue engineering, biosensing, and bioimaging over other 2D materials. All these applications systematically discussed in the following section.

3 Biomedical Applications of BP

3.1 Cancer Treatment

Cancer is a major life-threatening disease that can cause deaths more than those caused by AIDS, malaria, and tuberculosis combined [21]. Traditional standard approaches such as surgery, chemotherapy, and radiotherapy, etc., have been contributed greatly to reductions in cancer mortality for the past 50 years. However, all these modalities are highly invasive and frequently cause severe side effects due to the incomplete excision of the solid tumor as well as the lack of drug specificity towards tumorigenic cells. Therefore, advanced therapeutic strategies with high specificity, high efficacy, and low side effects have to be adopted for cancer treatment.

3.1.1 Phototherapy

In recent years, tumor-targeted "combination therapy" has attracted increasing attention especially in the field of nanomedicine which has proven more effective than single therapy in cancer treatment. This is mainly because; the therapy contributes to synergistic anti-cancer effects, low drug-related toxicity, and multidrug resistance inhibition via different mechanisms [31]. Phototherapy is one of the promising and non-invasive combination therapies which require radiation energy to eradicate cancer. In phototherapy, light energy converts into either chemical energy or heat energy which results in a series of chemical reactions in the body to reach the therapeutic effect [32]. Generally, phototherapy is classified into two types; are (i) photothermal therapy (PTT), and (ii) photodynamic therapy (PDT). BP has been proven as a potential substitute for conventional cancer therapies due to its negligible invasiveness and superior therapeutic efficacy. Furthermore, the remarkable carrier mobility and tunable band gap of BP leads to its broad absorption within UV and visible regions and hence BP acts as a potential photosensitizer to enhance PTT and PDT.

Photothermal therapy (PTT) In PTT, locally generated heating that induced by a photothermal agent will destruct the tumor cells under NIR light irradiation.

Ultrasmall BPQDs (~2.6 nm lateral size and 1.5 nm thickness) were synthesized via liquid exfoliation method and utilized as NIR photothermal agent [33]. The BPQDs have exhibited a large extinction coefficient of 14.8 $Lg^{-1}cm^{-1}$ at 808 nm, and photothermal conversion efficiency of about 28.4%, as well as good photostability. BPQDs stability in the physiological medium is improved by polyethylene glycol (PEG) surface modification and had negligible cytotoxicity to different types of cells.

The NIR photoexcitation of the BPQDs resulted in the significant death of C6 and MCF7 cancer cells suggesting that BPQDs are potential in PTT applications.

In another study, the efficiency of BP nanoparticles as a nanotheranostic agent in the detection and treatment of cancer via PTT has been assessed [34]. The water-soluble and PEGylated BP nanoparticles were synthesized by one-pot solventless high energy mechanical milling (HEMM) technique. The resultant nanoparticles have excellent biocompatibility and photostability and were able to convert NIR light into heat. The in vivo photoacoustic images indicate the efficient accumulation of BP nanoparticles via the enhanced permeability retention effect and these nanoparticles were utilized in photothermal ablation of cancer tumor under NIR irradiation. The results reveal that tumor-bearing mice were completely recovered after PTT and no significant toxicity was found in their normal tissues of main organs.

Even having good biocompatibility, and photothermal performance, the clinical applications of ultrasmall BPQDs are still restricted due to the deprivation of their optical properties in the aqueous medium and their fast renal excretion. To overcome this challenge, a core-shell type BPQD/PLGA nanospheres were processed using PLGA loaded with 3 nm BPQDs via an oil-in-water emulsion solvent evaporation method [22]. The hydrophobic nature of biodegradable and biocompatible PLGA polymer shell can isolate the internal BPQDs from the physiological environment and so that enhances the photothermal stability of BPQDs and controls their degradation rate. The in vitro and in vivo experiments demonstrate that the proposed nanospheres offer good bio-safety and tumor targeting ability and high efficiency of PTT under NIR laser illumination.

A covalent functionalization strategy was adopted to enhance the air and water stability of BPNSs [35]. Covalent modification of BPNSs was achieved using Nile Blue (NB) dye via aryl diazonium chemistry (Fig. 1a) and the produced NB@BPNSs exhibit both fluorescent imaging (Fig. 1b) and PTT (Fig. 1c) capabilities. In vitro experiments demonstrate that dye-modified BPNSs have good biocompatibility. The NB@BPNSs can mark the tumor site with red fluorescence and lead to efficient tumor ablation under NIR irradiation during in vivo tests (Fig. 1d).

BPNSs loaded cellulose based green and injectable composite hydrogels were developed via a chemical cross-linking reaction in alkaline medium [36]. The developed cellulose/BPNSs nanocomposite hydrogels with 3D network structure are completely biocompatible and non-toxic in both in vivo and in vitro experiments. The polymer hydrogels based photothermal agents exhibited excellent flexibility, notable photothermal response and showed great potential for PTT against cancer in mice.

A sprayable BP@PLEL hydrogel was adopted in a novel PTT system for postsurgical cancer therapy [37]. The sprayable hydrogel with superb NIR PTT performance exhibit a fast sol-gel transition and also had excellent biodegradability and biocompatibility in both in vivo and in vitro experiments. The small amount of sprayed hydrogel facilitates quick gelled membrane formation on wound under the irradiation with NIR light (0.5 W cm^{-2}) and exhibit strong PTT potential to remove the residual tumor tissues.

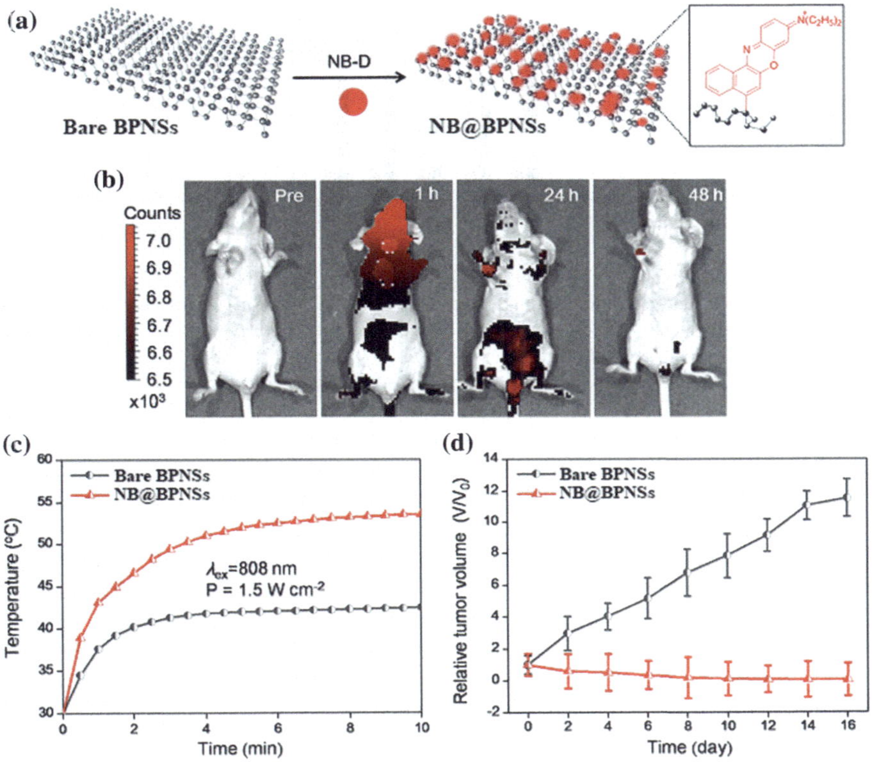

Fig. 1 **a** Schematic illustration of the construction of NB@BPNSs, **b** fluorescence images of the mice injected with NB@BPNSs at different time intervals, **c** time-dependent temperature increment, **d** changes in tumor volume after intravenous injection of BPNSs and NB@BPNSs (Reproduced with permission from American Chemical Society [35], Copyright 2017)

Photodynamic therapy (PDT) In PDT, a photosensitizer triggers a series of photochemical reactions under light irradiation, results in singlet oxygen and other reactive oxygen species (ROS), which can kill the targeted tumor cells. Owing to its long area of wavelength absorption, BP acts as a good photosensitizer and efficient for PDT.

In a study, water dispersible exfoliated BPNSs show notable singlet oxygen generation ability (Fig. 2a) with an elevated quantum yield of about 0.91, which is greater than other PDT agents [38]. The in vitro PDT effectiveness and cytotoxicity of BPNSs against tumor cells was evaluated by a model MTT viability assay, indicating that the ultrathin BP NSs revealed significant cell-growth inhibition (Fig. 2b). The intracellular ROS produced by BPNSs under light irradiation would induce cell apoptosis. The tumor growth in tumor-bearing mice was efficiently hampered by BPNSs under a small amount of light with short irradiation time (Fig. 2c). The therapeutic potential of BPNSs for the in vivo PDT of cancer was further highlighted by its rapid degradation into biocompatible phosphorous oxide without any residual.

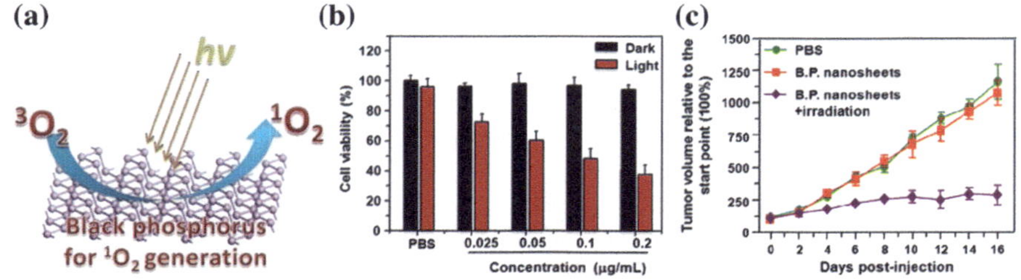

Fig. 2 **a** Schematic illustration of singlet oxygen production by BPNSs under laser irradiation, **b** effect of different concentrations of ultrathin BPNSs on cell viability, **c** time-dependent tumor growth after treatment with different formulations (Reproduced with permission from American Chemical Society [38], Copyright 2015)

Ultra small BPQDs with renal clearance properties were synthesized by a solvothermal method, and their stability and biocompatibility were enhanced via surface modification with PEG [39]. Cell viability examine was demonstrated that ultrathin BP QDs exhibited strong cytotoxicity against cancer cells under light irradiation. BPQDs which were incubated in the cancer cells produce the ROS under light irradiation, and the as-produced intracellular ROS were detected by a fluorescent dye (DCFH-DA). The ultrasmall BPQDs act as an efficient photosensitizer and exhibited excellent antitumor performance through PDT effect in S180 tumor-bearing BALB/c mice. Furthermore, the ultrasmall size (5.4 nm) of BPQDs facilitates their effective excretion from the body through renal clearance.

The unique surface structure of "Janus nanoparticles (JNPS)" with two different materials and chemistry properties open a new window to construct a nanohybrid platform for cancer therapeutics [40–42]. Based on this strategy, a new Janus nanoparticle (J-MOPs) was designed using BPQDs and tetrahydroxyanthraquinone (THQ)-Cu metal-organic particles (MOPs) via "host-metal-guest" coordination effects [43]. BPQDs were completely encapsulated by J-MOPs through P-Cu bonds and effectively isolated from the physiological environment. Additionally, J-MOPs could bestow the electron/hole separation and migration capability to BPQDs, which could improve the singlet oxygen production against cancer under 670 nm light irradiation. The discharged Cu^{2+} ions from the tumor-acidic activated J-MOPs degradation acts as Fenton-like agent to create OH from H_2O_2 which further enhance the photocatalytic activity of J-MOPs for PDT treatment of cancers in mice.

3.1.2 Therapeutic Agent (Drug) Delivery

Owing to its large surface area, BP exhibits higher drug loading capacity by combining the positively charged drug molecules with negatively charged phosphoric acid presents on its surface.

Chemotherapy is a compelling therapeutic approach in the field of cancer treatment [44, 45]. However, tumor cells with mutant p53 possess anti-apoptosis ability to tolerate chemotherapeutic drugs like doxorubicin (DOX) and hinder cancer therapy [46]. Phenethyl isothiocyanate (PEITC) can effectively deplete this intracellular mutant p53. In a study [47], multifunctional BP based drug delivery system was developed using BPNSs loaded with DOX (BPNSs/DOX) and coated with polydopamine (PDA) layer (BPNSs-PDA/DOX). The PDA layer was further reacted with PEG-$NH_2$to enhance the solubility and environmental stability of BPNSs. BPNSs-PDA-PEG/DOX was further decorated by PEITC via π-π stacking. The presence of PEITC inhibits the expression of mutant p53 and enhances the DOX therapeutic effect on human breast tumor cells. In addition, the phenolic hydroxyl groups of BPNSs-PDA-PEG-PEITC/DOX chelate with Mn^{2+} ions. The obtained BPNSs-PDA-PEG-PEITC-Mn/DOX utilized to study the biodistribution of nanocarriers through in vivo magnetic resonance imaging (MRI), and the results affirmed the efficient accumulation of nanocarriers at tumor locations. Combining those functions, BPNSs-PDA-PEG-PEITC-Mn/DOX could be used as a theranostic agent for magnetic MRI-guided PTT/PDT/chemotherapy of multiple drug resistant (MDR) cancers (Fig. 3).

In another investigation, a BPNSs based versatile drug delivery system (Fig. 4) was adopted for acute lymphoblastic leukemia (ALL) therapy [48]. BPNSs were prepared by liquid exfoliation method and their physiological stability was enhanced through surface modification with PEG. The processed PEGylated BPNSs (BPNSs@PEG) were efficient in photothermal conversion and exhibits excellent pH and NIR laser dual-responsive drug release property. The anti-cancer drug (DOX) molecules were loaded onto the BPNSs@PEG through electrostatic adsorption. ALL cells overexpress tyrosine kinase 7 (PTK7) and Sgc 8 aptamers have a strong fondness to PTK7. Therefore, to endow the nanocarriers with explicit targeting ability toward ALL cells, Sgc 8 aptamers were linked to the exterior of the BPNSs@PEG. The Sgc 8 aptamers function as the specific targeting ligand and due to the specific recognition between Sgc 8 aptamers and PTK7, the nanocarriers can be directed towards the ALL cells and execute specific drug delivery toward the target cells. Based on this strategy, one can obtain selectivity towards various types of tumors, by simply altering the targeting ligand on the nanocarriers.

Gene therapy has attracted considerable attention for diabetes, cardiovascular diseases, immunodeficiency diseases and cancers. However, the poor specificity, and low delivery and release efficiency of small interfering RNA (siRNA) restrict the clinical applications of gene therapy. BPQDs were utilized as a gene delivery platform for tumor healing, which offers effective delivery of lysine-specific dimethylase 1 (LSD1) siRNA into human ovarian teratocarcinoma PA-1 cells [49]. BPQDs were prepared by liquid phase exfoliation (LPE) technique and coated with polyelectrolyte polymer to develop BPQDs@PAH. The BPQDs-LSD1 siRNA complex showed greater transfection capability related to the commercial delivery reagents and effectively suppressed the expression of LSD1 mRNA inPA-1 cells. The combinatorial treatment of BPQDs-LSD1 siRNA complex with NIR laser inhibits the cell development rate by >80% with negligible cytotoxicity (Fig. 5).

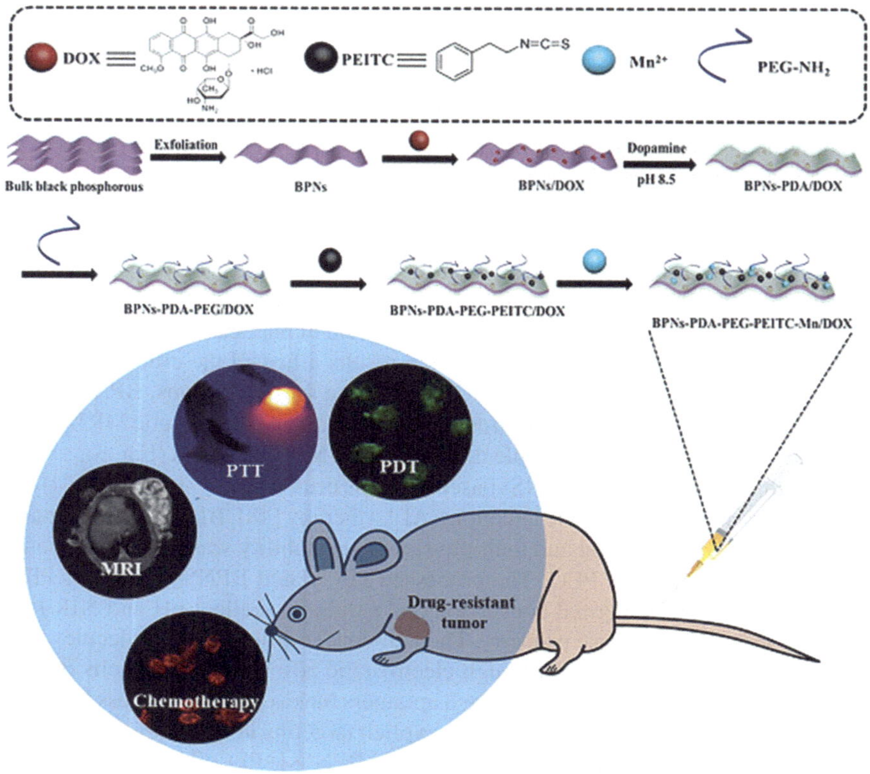

Fig. 3 Schematic illustration of the synthesis strategy and bioimaging-guided cancer therapy of BPNSs-PDA-PEG-PEITC-Mn/DOX (Reproduced with permission from Elsevier [47], Copyright 2019)

3.1.3 Applications in 3D Printing

Three-dimensional (3D) printing is also known as additive manufacturing technique, offers revolutionary research potentials in an expansive scope of medical applications including cell and tissue engineering [50, 51]. By utilizing advanced technology, 3D printing has emerged as an attractive area of research for the design and manufacturing of artificial scaffolds that are used for bone regeneration in modern oncological applications [52, 53]. A novel therapeutic platform was designed by coordinating2D BPNSs into 3D printed bioglass (BG) framework [54]. The developed bifunctional BP-BG framework can be utilized to heel osteosarcoma by PTT and in situ bone resurgence via phosphorous-driven, calcium extracted biomineralization process. In a stepwise treatment process; first, BPNSs employed as PTT agent due to their exceptional photothermal conversion capability. Next, due to their intrinsic properties with regard to osteogenesis, osteoconduction and osteoinduction, 3D printed BG frameworks play vital roles for the treatment of bone cancer. Finally, the scaffolds slowly

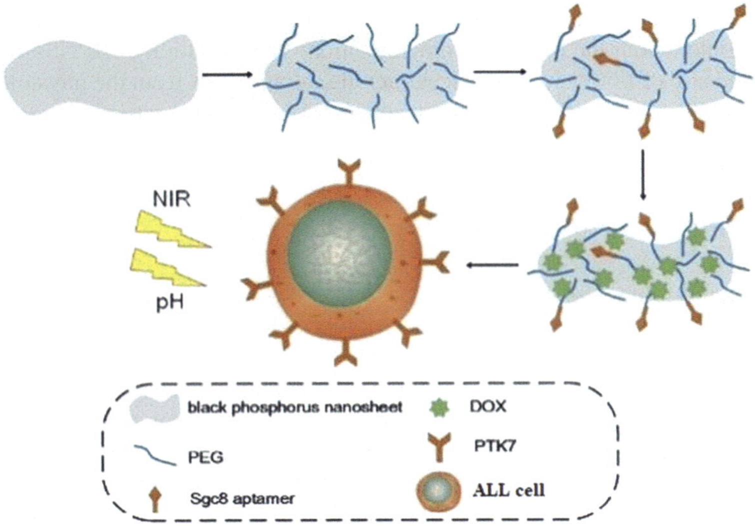

Fig. 4 Schematic illustration of the synthesis process of drug loaded and specific targeting ligand attached BPNSs@PEG for dual-responsive drug release (Reproduced with permission from American Chemical Society [48], Copyright 2019)

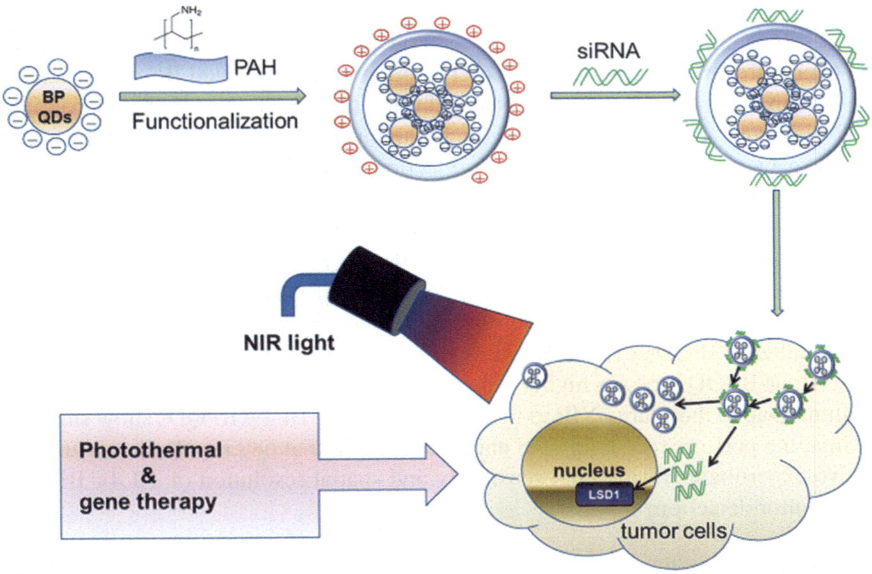

Fig. 5 Schematic illustration of a combination of photothermal and gene therapy by siRNA loaded BPQDs@PAH in cancer cells (Reproduced with permission from Royal Society of Chemistry [49], Copyright 2017)

degrade into new bone tissue components. The intrinsic physiochemical properties of BPNSs could favour the in situ biomineralization. The oxidation of BPNSs accelerates the release of PO_4^{3-} ions which extract the calcium ions from the physiological environment and create the calcium phosphate for bone regeneration.

3.2 Applications in Diagnosis

3.2.1 Bioimaging Agent

Photoacoustic imaging: Photoacoustic (PA) imaging is also known as optoacoustic or thermoacoustic imaging that exclusively combines the absorption contrast of light with ultrasound resolution to image human or animal organs with high contrast and high resolution [55, 56]. When nanodrug absorbs electromagnetic (EM) radiation energy, the absorbed energy transformed into heat and thus, the temperature of the nanodrug enhanced. Due to this temperature increment, thermal expansion takes place and generates acoustic force that behaviours the in vivo imaging of nanodrug based on ultrasonic signals [57]. Considering its photoacoustic signals and exceptional photothermal response, BP can be suitable for internal photoacoustic imaging.

Water-soluble PEGylated BP nanoparticles were prepared by solventless high energy mechanical milling (HEMM) method and were utilized for PA imaging and cancer treatment applications [34]. The processed PEGylated BP nanoparticles exhibit capability of conversion of NIR light into heat and this could be appropriate for PA imaging. The tumor retains the higher intensity of signal than that of liver and kidney, even after 24 h of intravenous injection with PEGylated BP nanoparticles (Fig. 6). This suggests the accumulation of a large number of BP nanoparticles in the tumor site through the influence of enhanced permeability retention (EPR) effect and the easy excretion of nanoparticles from the liver and kidney.

Surface modification of BPQDs via coordination with a sulfonic ester of the titanium ligand (TiL_4) promotes their enhanced stability in aqueous medium and the functionalized BPQDs (TiL_4@BPQDs) were employed for PA imaging of cancer [58]. TiL_4@BPQDs possess higher stability than that of bare BPQDs in physiological medium. Due to their large NIR extinction coefficient, TiL_4@BPQDs show superior PA imaging performance than gold nanorods (AuNRs) at 680 nm. Both in vitro and in vivo experiments reveal high sensitivity and spatial resolution of TiL4@BPQDs in the tumor detection.

Fluorescence imaging Fluorescence imaging in the NIR region is a low cost and high sensitivity method of real-time molecular imaging in vivo that has enabled a large variety of clinical technologies [59].

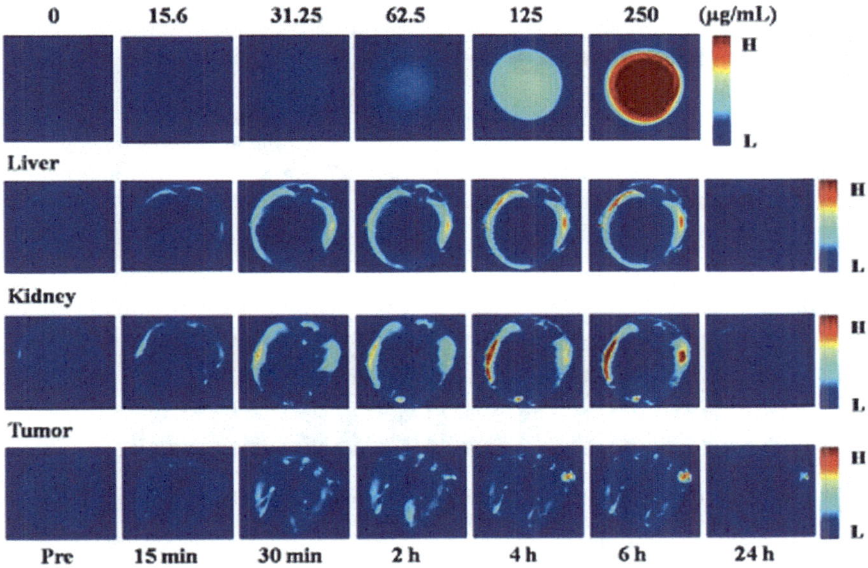

Fig. 6 In-vitro photo acoustic images of PEGylated BP nanoparticles solution (first row), and in vivo PA images of liver, kidney, and tumor at different times (Reproduced with permission from Elsevier [34], Copyright 2016)

Due to its layer dependant fluorescence spectrum, few-layered BP able to emit fluorescence and exhibit a peak with high intensity at 775 nm. In a study [60], ultra-small BPQDs (~3.6 nm) were prepared via a liquid exfoliation method and assembled with magnetic iron oxide carbon dots (Fe_3O_4-CDs) to construct the genipin (GP)-wrapped polyglutamic acid (PGA)-Fe_3O_4-CDs@BPQDs nanocomposite for fluorescence imaging-guided phototherapy of cancer (Fig. 7a). After injection with GP-PGA-Fe_3O_4-CDs@BPQDs, the HeLa tumor-bearing mice exhibit strong fluorescence semaphore at the tumor area (Fig. 7b). The GP-PGA-Fe_3O_4-CDs@BPQDs accumulated in the tumor area through EPR effect and thus fluorescence was detected in the tumor site. Clinical analyses reveal that nanocomposite did not show any side effects in mice.

Due to its large specific surface area, BP carried with fluorescent molecules for the in vivo fluorescence imaging. Chlorin e6 (Ce6) is a commercial photosensitizer which has been extensively employed for NIR fluorescence imaging [61]. BPNSs were fabricated via liquid exfoliation method and modified with PEG-NH_2 for their biocompatibility enhancement [62]. The processed BP@PEG NSs are further loaded with Ce6 to develop BP@PEG/Ce6 NSs which exhibit good biocompatibility, aqueous medium stability, and tumor-targeting ability. In vivo fluorescence imaging demonstrates the effective accumulation of BP@PEG/Ce6 NSs in the tumor through the influence of EPR effect. The nanosystem exhibits excellent potential for fluorescence/NIR imaging-guided phototherapy in the clinic.

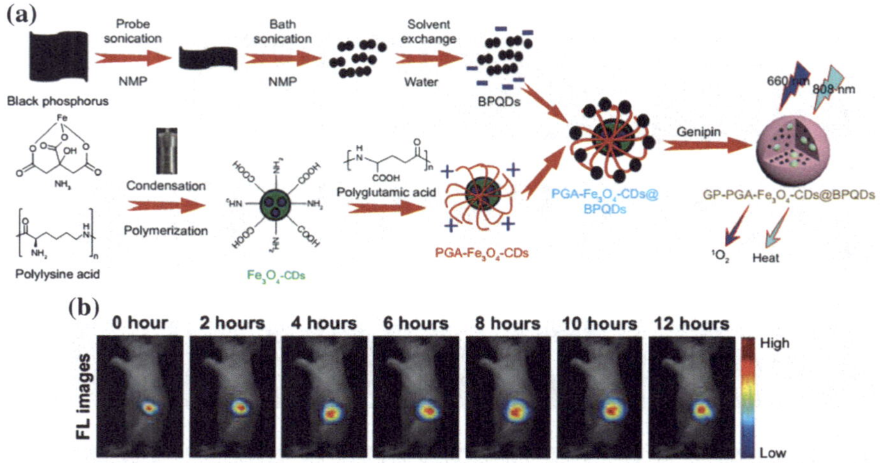

Fig. 7 **a** Preparation of GP-PGA-Fe_3O_4-CDs@BPQDs nanocomposite, **b** real-time in vivo fluorescence imaging with different post-injection times (Reproduced with permission from Dove Media Press Limited [60], Copyright 2018)

Surface-enhanced Raman scattering (SERS) imaging SERS imaging is a powerful and label-free analytical technique for biomarker detection, intracellular tracking and pathological diagnosis [63].

In a study [64], BPNSs were prepared through liquid exfoliation method and loaded with gold (Au) nanoparticles through the in situ chemical reduction of gold salts on BPNSs (Fig. 8). The developed BP-AuNPs possess excellent photostability with enhanced NIR absorption. The high SERS activity of BP-AuNPs under NIR

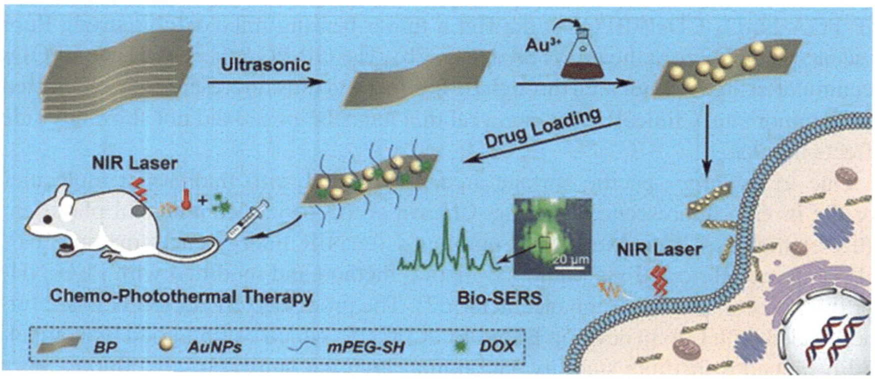

Fig. 8 Schematic presentation of the synthesis of BP-AuNPs hybrids and their applications in bio-SERS (Reproduced with permission from Royal Society of Chemistry [64], Copyright 2018)

laser irradiation with a low inherent background, endow them as promising molecular fingerprint probe for SERS biological analysis. The biocomponent identification, tumor cell classification, tracking of endocytosed nanostructure and real-time monitoring of photothermal ablation of tumor cells have been attained by the utilization of BP-AuNPs as a SERS substrate.

3.2.2 Biosensing

Owing to its fascinating properties including, anisotropic electrical conductivity, photoresponse, high carrier mobility, and tunable band gap, BP can be considered as a potential material for the construction of electronics and optoelectronics devices for molecular sensing.

Poly-L-lysine (pLL) is a cationic polymer and frequently employed for biomolecules sensing. In a study [65], pLL adhered to the surface of BP nanoflakes through electrostatic (hydrophobic) interactions to construct the pLL-Bp hybrid for hydrogen peroxide (H_2O_2) sensing (Fig. 9a). A negatively charged protein hemoglobin (Hb) was immobilized on to the surface of pLL-Bp hybrid and the resulting Hb@pLL-Bp hybrid acts as an enzymatic electrochemical biosensor and exhibits potential biocatalytic performance towards the reduction of oxygen and

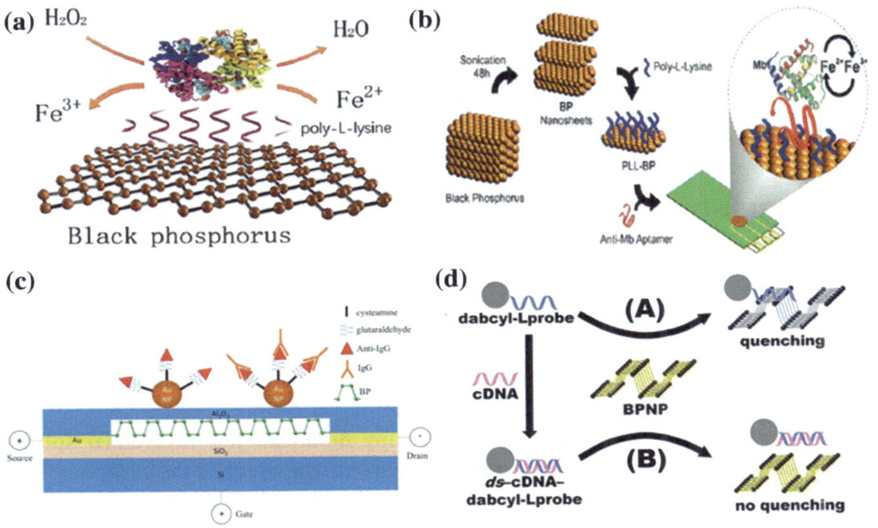

Fig. 9 **a** Schematic of BP for H_2O_2 sensing (Reproduced with permission from American Chemical Society [65], Copyright 2018), **b** schematic of surface modification of BP for Mb detection (Reproduced with permission from American Chemical Society [66], Copyright 2016), **c** schematic of a BP FET biosensor for IgG detection (Reproduced with permission from Elsevier [28], Copyright 2017), **d** schematic of fluorescent detection of DNA using BPNPs as fluorophores (Reproduced with permission from Royal Society of Chemistry [67], Copyright 2017)

H_2O_2. The results indicate that pLL-Bp hybrid function as an effective matrix for fast direct electron transfer (DET) and thus useful for biosensors and bioelectronics applications.

The pLL-Bp hybrid also applied for the electrochemical diagnosis of the redox active cardiac biomarker myoglobin (Mb) in serum samples [66]. The pLL acts as a linker for further biomolecular interactions through the nitrogen atoms. The Coulombic interactions between pLL and DNA facilitate the immobilization of negatively charged DNA aptamers for Mb onto the BPNSs. Electron transfer mechanism could endow the direct oxidation of Mb to Fe^{3+} at the nanostructured electrode surface (Fig. 9b). The aptamer functionalized pLL-Bp hybrid exhibit a very small diagnosis limit (0.524 pg mL^{-1}) and sensitivity (36 μA pg^{-1} mL cm^{-2}) for myoglobin.

A surface passivated field-effect transistor (FET) biosensor was constructed for human immunoglobulin G (IgG) detection [28] (Fig. 9c). Few layered BP nanosheets were fabricated by mechanical exfoliation method and coated with an Al_2O_3 thin film for surface passivation by using atomic layer deposition. Gold nanoparticles conjugated with anti-human IgG were deposited on the surface of BP device as the probes for antigen (IgG) detection. The as-produced BP device exhibits a high sensitivity to human IgG of about 10 ng $mL^{-1}$with quick response time, suggests its excellent biosensing capability for rapid diagnostics of diseases.

Due to its fluorescent nature, BP employed as an efficient fluorescent quencher in fluorescence-based biosensing. In a study, BPNPs utilized as nanofluorophores for DNA diagnosis [67] (Fig. 9d). The as-constructed BP fluorescent platform exhibits a linear detection range from 4 pM to 4000 pM with good linearity of about r of 0.91. A high sensitivity of DNA detection (5.9 pM) with a low limit of quantification (19.7 pM) was achieved by the proposed platform.

4 Theranostics

Theranostics is a new field of medicine, which combines diagnostics and therapy, thus enabling imaging-guided therapy. Owing to its astounding optical property, biocompatibility, and biodegradability, BP play role as promising theranostic agent for in vitro and in vivo imaging, with PTT and PDT features for cancer treatment [68].

Despite its excellent potential in cancer therapy, mono and few-layered BP suffers from rapid degradation in the physiological environment which is the Achilles' heel of BP, hinders its practical applications in cancer theranostics. Polydopamine (PDA) coated BPQDs were developed through a self-polymerization method which provides an easy surface passivation system to overcome the Achilles' heel of BP [69]. PDA with improved phenol groups is a scavenger of ROS which prevents the oxidation BPQDs from oxygen and moisture and enhances their stability in an aqueous medium. Furthermore, PDA is an excellent photothermal material with strong NIR absorption could enhance the photothermal conversion efficiency (PCE) of BPQDs from 22.6 to 64.2%. Therefore, BP@PDA exhibit greater potential for photoacoustic imaging

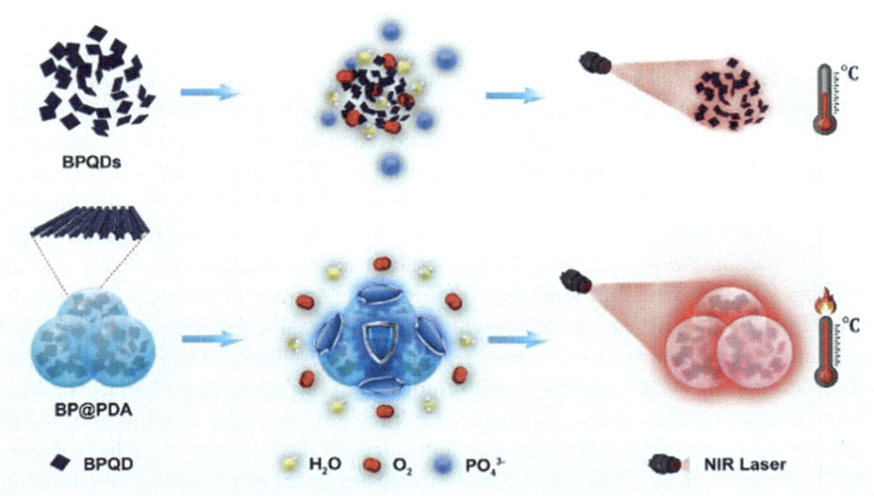

Fig. 10 Schematic illustration of BP@PDA as a cancer theranostic agent with excellent stability and PCE compared to BPQDs (Reproduced with permission from Elsevier [69], Copyright 2019)

(PAI) guided PTT of cancer theranostics as compared to BPQDs, which has been examined via in vitro and in vivo experiments under 808 nm laser (Fig. 10).

The conventional ultraviolet/visible light has low tissue penetration which hinders the medical applications of PDT for cancer treatment. Furthermore, BP based theranostics platform requires two different light sources for real-time diagnostic applications. Therefore, the construction of a theranostics agent having therapeutic and diagnostic functions under a single irradiation light is still a major challenge in PDT [70]. Up-conversion luminescent (UCL) materials can effectively translate the NIR light to visible light via multiple-photon process, which has been applied in imaging, cancer therapy, and drug delivery fields [71, 72]

In an investigation, a multifunctional composite was constructed through the combination of integrating up-conversion nanoparticles (UCNPs) and BPNSs to achieve PDT under a single 808 nm laser light excitation [73]. Initially, the $NaGdF_4$:Yb,Er@Yb@Nd@Yb UCNPs were tailored with poly(acrylic acid) and then integrated with PEGylated BPNSs via electrostatic interactions. The processed UCNP@BP exhibited greater anti-tumor effect which has been examined through in vitro and in vivo experiments under 808 nm light irradiation.

5 Summary and Future Perspectives

This chapter reviewed some of the important literature on black phosphorous (BP) to emphasize the modern research in cancer therapy, biosensing, and imaging biomedical applications. Due to its fascinating properties such as, tunable band gap, functional anisotropy, large surface area, and biocompatibility; BP plays a vital role in a broad range of biomedical applications. The physiological stability of BP was enhanced via surface chemical modifications and surface protective layer construction. In spite of the remarkable advancement, the exploration in this domain still meets several challenges in terms of synthetic methodology, toxicity and stability. Bulk production of high-quality BP nanoparticles which is indispensable for potential biomedical applications with minimal cytotoxicity remains a big challenge. The impact of BP on cell functionalities (e.g., viability, proliferation, and differentiation) and its medical diagnosis applications have to be explored in future studies. The combination of BP with other alternative 2D materials (e.g., antimonene QDs) to construct a hybrid therapeutic platform would be expected as an interesting area of research in the coming years for advanced cancer therapy.

References

1. de Menezes, B.R.C., Rodrigues, K.F., da Silva Fonseca, B.C., et al.: Recent advances in the use of carbon nanotubes as smart biomaterials. J. Mater. Chem. B **7**, 1343–1360 (2019). https://doi.org/10.1039/C8TB02419G
2. Patz, J.A., Graczyk, T.K., Geller, N., Vittor, A.Y.: Effects of environmental change on emerging parasitic diseases. Int. J. Parasitol. **30**, 1395–1405 (2000). https://doi.org/10.1016/S0020-7519(00)00141-7
3. Kumar, S., Rani, R., Dilbaghi, N., et al.: Carbon nanotubes: a novel material for multifaceted applications in human healthcare. Chem. Soc. Rev. **46**, 158–196 (2017). https://doi.org/10.1039/C6CS00517A
4. Partha, R., Conyers, J.L.: Biomedical applications of functionalized fullerene-based nanomaterials. Int. J. Nanomed. **4**, 261–275 (2009)
5. Gao, J., Wang, H.L., Shreve, A., Iyer, R.: Fullerene derivatives induce premature senescence: a new toxicity paradigm or novel biomedical applications. Toxicol. Appl. Pharmacol. **244**, 130–143 (2010). https://doi.org/10.1016/J.TAAP.2009.12.025
6. Erol, O., Uyan, I., Hatip, M., et al.: Recent advances in bioactive 1D and 2D carbon nanomaterials for biomedical applications. Nanomed. Nanotechnol. Biol. Med. **14**, 2433–2454 (2018). https://doi.org/10.1016/J.NANO.2017.03.021
7. Bianco, A., Kostarelos, K., Partidos, C.D., Prato, M.: Biomedical applications of functionalised carbon nanotubes. Chem. Commun. **0**, 571 (2005). https://doi.org/10.1039/b410943k
8. Klingeler, R., Sim, R.B.: Carbon Nanotubes for Biomedical Applications. Springer, Berlin, Heidelberg (2011)
9. Shen, H., Zhang, L., Liu, M., Zhang, Z.: Biomedical applications of graphene. Theranostics **2**, 283–294 (2012). https://doi.org/10.7150/thno.3642
10. Yang, Y., Asiri, A.M., Tang, Z., et al.: Graphene based materials for biomedical applications. Mater. Today **16**, 365–373 (2013). https://doi.org/10.1016/J.MATTOD.2013.09.004
11. Zhang, Y., Nayak, T.R., Hong, H., Cai, W.: Graphene: a versatile nanoplatform for biomedical applications. Nanoscale **4**, 3833 (2012). https://doi.org/10.1039/c2nr31040f

12. Chimene, D., Alge, D.L., Gaharwar, A.K.: Two-dimensional nanomaterials for biomedical applications: emerging trends and future prospects. Adv. Mater. **27**, 7261–7284 (2015). https://doi.org/10.1002/adma.201502422
13. Luo, M., Fan, T., Zhou, Y., et al.: 2D black phosphorus-based biomedical applications. Adv. Funct. Mater. **29**, 1808306 (2019). https://doi.org/10.1002/adfm.201808306
14. Li, L., Yu, Y., Ye, G.J., et al.: Black phosphorus field-effect transistors. Nat. Nanotechnol. **9**, 372–377 (2014). https://doi.org/10.1038/nnano.2014.35
15. Liu, H., Neal, A.T., Zhu, Z., et al.: Phosphorene: an unexplored 2D semiconductor with a high hole mobility. ACS Nano **8**, 4033–4041 (2014). https://doi.org/10.1021/nn501226z
16. Castellanos-Gomez, A., Vicarelli, L., Prada, E., et al.: Isolation and characterization of few-layer black phosphorus. 2D Mater. **1**, 025001 (2014). https://doi.org/10.1088/2053-1583/1/2/025001
17. Wang, X., Jones, A.M., Seyler, K.L., et al.: Highly anisotropic and robust excitons in mono-layer black phosphorus. Nat. Nanotechnol. **10**, 517–521 (2015). https://doi.org/10.1038/nnano.2015.71
18. Wang, Y.W., Liu, S., Zeng, B.W., et al.: Ultraviolet saturable absorption and ultrafast carrier dynamics in ultrasmall black phosphorus quantum dots. Nanoscale **9**, 4683–4690 (2017). https://doi.org/10.1039/C6NR09235G
19. Wu, J., Koon, G.K.W., Xiang, D., et al.: Colossal ultraviolet photoresponsivity of few-layer black phosphorus. ACS Nano **9**, 8070–8077 (2015). https://doi.org/10.1021/acsnano.5b01922
20. Zhu, M., Osakada, Y., Kim, S., et al.: Black phosphorus: a promising two dimensional visible and near-infrared-activated photocatalyst for hydrogen evolution. Appl. Catal. B Environ. **217**, 285–292 (2017). https://doi.org/10.1016/J.APCATB.2017.06.002
21. Qiu, M., Wang, D., Liang, W., et al.: Novel concept of the smart NIR-light–controlled drug release of black phosphorus nanostructure for cancer therapy. Proc. Natl. Acad. Sci. **115**, 501–506 (2018). https://doi.org/10.1073/PNAS.1714421115
22. Shao, J., Xie, H., Huang, H., et al.: Biodegradable black phosphorus-based nanospheres for in vivo photothermal cancer therapy. Nat. Commun. **7**, 12967 (2016). https://doi.org/10.1038/ncomms12967
23. Fojtů, M., Chia, X., Sofer, Z., et al.: Black phosphorus nanoparticles potentiate the anticancer effect of oxaliplatin in ovarian cancer cell line. Adv. Funct. Mater. **27**, 1701955 (2017). https://doi.org/10.1002/adfm.201701955
24. Lee, H.U., Park, S.Y., Lee, S.C., et al.: Black phosphorus (BP) nanodots for potential biomedical applications. Small **12**, 214–219 (2016). https://doi.org/10.1002/smll.201502756
25. Wei, Q., Peng, X.: Superior mechanical flexibility of phosphorene and few-layer black phosphorus. Appl. Phys. Lett. **104**, 251915 (2014). https://doi.org/10.1063/1.4885215
26. Jiang, J.-W., Park, H.S.: Mechanical properties of single-layer black phosphorus. J. Phys. D Appl. Phys. **47**, 385304 (2014). https://doi.org/10.1088/0022-3727/47/38/385304
27. Li, X., Deng, B., Wang, X., et al.: Synthesis of thin-film black phosphorus on a flexible substrate. 2D Mater. **2**, 031002 (2015). https://doi.org/10.1088/2053-1583/2/3/031002
28. Chen, Y., Ren, R., Pu, H., et al.: Field-effect transistor biosensors with two-dimensional black phosphorus nanosheets. Biosens. Bioelectron. **89**, 505–510 (2017). https://doi.org/10.1016/J.BIOS.2016.03.059
29. Zhang, X., Zhang, Z., Zhang, S., et al.: Size effect on the cytotoxicity of layered black phosphorus and underlying mechanisms. Small **13**, 1701210 (2017). https://doi.org/10.1002/smll.201701210
30. Latiff, N.M., Teo, W.Z., Sofer, Z., et al.: The cytotoxicity of layered black phosphorus. Chem. A Eur. J. **21**, 13991–13995 (2015). https://doi.org/10.1002/chem.201502006
31. Lin, J., Wang, S., Huang, P., et al.: Photosensitizer-loaded gold vesicles with strong plasmonic coupling effect for imaging-guided photothermal/photodynamic therapy. ACS Nano **7**, 5320–5329 (2013). https://doi.org/10.1021/nn4011686
32. Thangavel, S., Yoshitomi, T., Sakharkar, M.K., Nagasaki, Y.: Redox nanoparticle increases the chemotherapeutic efficiency of pioglitazone and suppresses its toxic side effects. Biomaterials **99**, 109–123 (2016). https://doi.org/10.1016/J.BIOMATERIALS.2016.05.001

33. Sun, Z., Xie, H., Tang, S., et al.: Ultrasmall black phosphorus quantum dots: synthesis and use as photothermal agents. Angew. Chemie. Int. Ed. **54**, 11526–11530 (2015). https://doi.org/10.1002/anie.201506154
34. Sun, C., Wen, L., Zeng, J., et al.: One-pot solventless preparation of PEGylated black phosphorus nanoparticles for photoacoustic imaging and photothermal therapy of cancer. Biomaterials **91**, 81–89 (2016). https://doi.org/10.1016/J.BIOMATERIALS.2016.03.022
35. Zhao, Y., Tong, L., Li, Z., et al.: Stable and multifunctional dye-modified black phosphorus nanosheets for near-infrared imaging-guided photothermal therapy. Chem. Mater. **29**, 7131–7139 (2017). https://doi.org/10.1021/acs.chemmater.7b01106
36. Xing, C., Chen, S., Qiu, M., et al.: Conceptually novel black phosphorus/cellulose hydrogels as promising photothermal agents for effective cancer therapy. Adv. Healthc. Mater. **7**, 1701510 (2018). https://doi.org/10.1002/adhm.201701510
37. Shao, J., Ruan, C., Xie, H., et al.: Black-phosphorus-incorporated hydrogel as a sprayable and biodegradable photothermal platform for postsurgical treatment of cancer. Adv. Sci. **5**, 1700848 (2018). https://doi.org/10.1002/advs.201700848
38. Wang, H., Yang, X., Shao, W., et al.: Ultrathin black phosphorus nanosheets for efficient singlet oxygen generation. J. Am. Chem. Soc. **137**, 11376–11382 (2015). https://doi.org/10.1021/jacs.5b06025
39. Guo, T., Wu, Y., Lin, Y., et al.: Black phosphorus quantum dots with renal clearance property for efficient photodynamic therapy. Small **14**, 1702815 (2018). https://doi.org/10.1002/smll.201702815
40. Song, G., Chen, M., Zhang, Y., et al.: Janus iron oxides @ semiconducting polymer nanoparticle tracer for cell tracking by magnetic particle imaging. Nano Lett. **18**, 182–189 (2018). https://doi.org/10.1021/acs.nanolett.7b03829
41. Ju, Y., Zhang, H., Yu, J., et al.: Monodisperse Au–Fe_2C Janus nanoparticles: an attractive multifunctional material for triple-modal imaging-guided tumor photothermal therapy. ACS Nano **11**, 9239–9248 (2017). https://doi.org/10.1021/acsnano.7b04461
42. Song, J., Wu, B., Zhou, Z., et al.: Double-layered plasmonic-magnetic vesicles by self-assembly of Janus amphiphilic gold-iron(II, III) oxide nanoparticles. Angew. Chemie. Int. Ed. **56**, 8110–8114 (2017). https://doi.org/10.1002/anie.201702572
43. Zhang, D., Lin, Z., Lan, S., et al.: The design of Janus black phosphorus quantum dots@metal–organic nanoparticles for simultaneously enhancing environmental stability and photodynamic therapy efficiency. Mater. Chem. Front. **3**, 656–663 (2019). https://doi.org/10.1039/C8QM00623G
44. Dong, Z., Gong, H., Gao, M., et al.: polydopamine nanoparticles as a versatile molecular loading platform to enable imaging-guided cancer combination therapy. Theranostics **6**, 1031–1042 (2016). https://doi.org/10.7150/thno.14431
45. Sun, X., Wang, C., Gao, M., et al.: Remotely controlled red blood cell carriers for cancer targeting and near-infrared light-triggered drug release in combined photothermal-chemotherapy. Adv. Funct. Mater. **25**, 2386–2394 (2015). https://doi.org/10.1002/adfm.201500061
46. Ogretmen, B., Safa, A.R.: Expression of the mutated p53 tumor suppressor protein and its molecular and biochemical characterization in multidrug resistant MCF-7/Adr human breast cancer cells. Oncogene **14**, 499–506 (1997). https://doi.org/10.1038/sj.onc.1200855
47. Wu, F., Zhang, M., Chu, X., et al.: Black phosphorus nanosheets-based nanocarriers for enhancing chemotherapy drug sensitiveness via depleting mutant p53 and resistant cancer multimodal therapy. Chem. Eng. J. **370**, 387–399 (2019). https://doi.org/10.1016/J.CEJ.2019.03.228
48. Zong, S., Wang, L., Yang, Z., et al.: Black phosphorus-based drug nanocarrier for targeted and synergetic chemophotothermal therapy of acute lymphoblastic leukemia. ACS Appl. Mater. Interfaces **11**, 5896–5902 (2019). https://doi.org/10.1021/acsami.8b22563
49. Yin, F., Hu, K., Chen, S., et al.: Black phosphorus quantum dot based novel siRNA delivery systems in human pluripotent teratoma PA-1 cells. J. Mater. Chem. B **5**, 5433–5440 (2017). https://doi.org/10.1039/C7TB01068K
50. Baumann, B., Jungst, T., Stichler, S., et al.: Control of nanoparticle release kinetics from 3D printed hydrogel scaffolds. Angew. Chemie. Int. Ed. **56**, 4623–4628 (2017). https://doi.org/10.1002/anie.201700153

51. Khademhosseini, A., Langer, R.: A decade of progress in tissue engineering. Nat. Protoc. **11**, 1775–1781 (2016). https://doi.org/10.1038/nprot.2016.123
52. Zhang, Y., Zhai, D., Xu, M., et al.: 3D-printed bioceramic scaffolds with antibacterial and osteogenic activity. Biofabrication **9**, 025037 (2017). https://doi.org/10.1088/1758-5090/aa6ed6
53. Shi, W., Sun, M., Hu, X., et al.: Structurally and functionally optimized silk-fibroin-gelatin scaffold using 3D printing to repair cartilage injury in vitro and in vivo. Adv. Mater. **29**, 1701089 (2017). https://doi.org/10.1002/adma.201701089
54. Yang, B., Yin, J., Chen, Y., et al.: 2D-black-phosphorus-reinforced 3D-printed scaffolds: a stepwise countermeasure for osteosarcoma. Adv. Mater. **30**, 1705611 (2018). https://doi.org/10.1002/adma.201705611
55. Xu, M., Wang, L.V.: Photoacoustic imaging in biomedicine. Rev. Sci. Instrum. **77**, 041101 (2006). https://doi.org/10.1063/1.2195024
56. Li, C., Wang, L.V.: Photoacoustic tomography and sensing in biomedicine. Phys. Med. Biol. **54**, R59–R97 (2009). https://doi.org/10.1088/0031-9155/54/19/R01
57. Wang, L.V.: Multiscale photoacoustic microscopy and computed tomography. Nat. Photon. **3**, 503–509 (2009). https://doi.org/10.1038/nphoton.2009.157
58. Sun, Z., Zhao, Y., Li, Z., et al.: TiL_4—coordinated black phosphorus quantum dots as an efficient contrast agent for in vivo photoacoustic imaging of cancer. Small **13**, 1602896 (2017). https://doi.org/10.1002/smll.201602896
59. Carr, J.A., Franke, D., Caram, J.R., et al.: Shortwave infrared fluorescence imaging with the clinically approved near-infrared dye indocyanine green. Proc. Natl. Acad. Sci. **115**, 4465–4470 (2018). https://doi.org/10.1073/PNAS.1718917115
60. Zhang, M., Wang, W., Cui, Y., et al.: Near-infrared light-mediated photodynamic/photothermal therapy nanoplatform by the assembly of Fe_3O_4 carbon dots with graphitic black phosphorus quantum dots. Int. J. Nanomed. **13**, 2803–2819 (2018). https://doi.org/10.2147/IJN.S156434
61. Liu, T., Wang, C., Cui, W., et al.: Combined photothermal and photodynamic therapy delivered by PEGylated MoS_2 nanosheets. Nanoscale **6**, 11219–11225 (2014). https://doi.org/10.1039/C4NR03753G
62. Yang, X., Wang, D., Shi, Y., et al.: Black phosphorus nanosheets immobilizing Ce6 for imaging-guided photothermal/photodynamic cancer therapy. ACS Appl. Mater. Interfaces **10**, 12431–12440 (2018). https://doi.org/10.1021/acsami.8b00276
63. Lane, L.A., Qian, X., Nie, S.: SERS nanoparticles in medicine: from label-free detection to spectroscopic tagging. Chem. Rev. **115**, 10489–10529 (2015). https://doi.org/10.1021/acs.chemrev.5b00265
64. Liu, Z., Chen, H., Jia, Y., et al.: A two-dimensional fingerprint nanoprobe based on black phosphorus for bio-SERS analysis and chemo-photothermal therapy. Nanoscale **10**, 18795–18804 (2018). https://doi.org/10.1039/C8NR05300F
65. Zhao, Y., Zhang, Y.-H., Zhuge, Z., et al.: Synthesis of a poly-l-lysine/black phosphorus hybrid for biosensors. Anal. Chem. **90**, 3149–3155 (2018). https://doi.org/10.1021/acs.analchem.7b04395
66. Kumar, V., Brent, J.R., Shorie, M., et al.: Nanostructured aptamer-functionalized black phosphorus sensing platform for label-free detection of myoglobin, a cardiovascular disease biomarker. ACS Appl. Mater. Interfaces **8**, 22860–22868 (2016). https://doi.org/10.1021/acsami.6b06488
67. Yew, Y.T., Sofer, Z., Mayorga-Martinez, C.C., Pumera, M.: Black phosphorus nanoparticles as a novel fluorescent sensing platform for nucleic acid detection. Mater. Chem. Front. **1**, 1130–1136 (2017). https://doi.org/10.1039/C6QM00341A
68. Qiu, M., Singh, A., Wang, D., et al.: Biocompatible and biodegradable inorganic nanostructures for nanomedicine: silicon and black phosphorus. Nano Today **25**, 135–155 (2019). https://doi.org/10.1016/J.NANTOD.2019.02.012
69. Li, Z., Xu, H., Shao, J., et al.: Polydopamine-functionalized black phosphorus quantum dots for cancer theranostics. Appl. Mater. Today **15**, 297–304 (2019). https://doi.org/10.1016/J.APMT.2019.02.002

70. Li, Y., Tang, J., Pan, D.-X., et al.: A versatile imaging and therapeutic platform based on dual-band luminescent lanthanide nanoparticles toward tumor metastasis inhibition. ACS Nano **10**, 2766–2773 (2016). https://doi.org/10.1021/acsnano.5b07873
71. Chen, G., Ågren, H., Ohulchanskyy, T.Y., Prasad, P.N.: Light upconverting core–shell nanostructures: nanophotonic control for emerging applications. Chem. Soc. Rev. **44**, 1680–1713 (2015). https://doi.org/10.1039/C4CS00170B
72. Zheng, W., Huang, P., Tu, D., et al.: Lanthanide-doped upconversion nano-bioprobes: electronic structures, optical properties, and biodetection. Chem. Soc. Rev. **44**, 1379–1415 (2015). https://doi.org/10.1039/C4CS00178H
73. Lv, R., Yang, D., Yang, P., et al.: Integration of upconversion nanoparticles and ultra-thin black phosphorus for efficient photodynamic theranostics under 808 nm near-infrared light irradiation. Chem. Mater. **28**, 4724–4734 (2016). https://doi.org/10.1021/acs.chemmater.6b01720

Structure and Fundamental Properties of Black Phosphorus

Mohd Imran Ahamed, Nimra Shakeel and Naushad Anwar

Abstract Black phosphorus (BP) is one of the most stable allotropes among the three allotropes of phosphorus at a high temperature under a high pressure possessing new two-dimensional layered structure, which was first prepared by Bridgman in 1914. Since, the development and recent success in growing two-dimensional material family, single- or few-layered, BP has recently a vital field of interest due to their various superior properties, such as a tunable and direct/narrow band gaps, high carrier mobility, large specific surface area, photothermal property, biocompatibility, biodegradability and many interesting in-layer anisotropies and attracted considerable attention on applications in energy conversion and storage, oxygen evolution, electronics, optoelectronics, photocatalytic hydrogenation, water splitting, and thermoelectric generators, etc. Especially, there emerge contributions on electrochemical energy storage devices as supercapacitors and in batteries like lithium/sodium ion batteries. This chapter summarizes the structure and fundamental properties and few preparation methods of BP.

Keywords Black phosphorus · Structural and hybridization · Fundamental properties · Preparation

1 Introduction

Phosphorus is one of the essential elements to maintain good health in the human body. The major allotropes of phosphorus are white phosphorus (WP), red phosphorus (RP) and black Phosphorus (BP) [1, 2]. WP is chemically unstable due to large bond strain and possesses tetrahedral diamond like structure. RP possesses an irregular structure like an amorphous solid and more chemically stable as compare to WP [3, 4]. In contrast, at ambient conditions, BP is the most stable and is nonflammable and insoluble in most solvents. These allotropes are mutually transformed to one another under certain conditions such as temperature and pressure [5, 6]. Bridgman first studied the structure of BP in 1914. BP was accidentally obtained

M. I. Ahamed · N. Shakeel · N. Anwar (✉)
Department of Chemistry, Aligarh Muslim University, Aligarh 202002, U.P., India
e-mail: naushadanwar85@gmail.com

Inamuddin et al. (eds.), *Black Phosphorus*, Engineering Materials,
https://doi.org/10.1007/978-3-030-29555-4_7

by Bridgeman when he applied the effect of high pressure on the melting point of WP. He also studied on the fundamental properties and measured the specific density (2.69 g cm^{-3}), magnetic permeability, specific heat, etc. [7]. Another study on the crystal structure of BP was put forward by Hultgren et al. [8] and further investigated by him in 1935 using X-ray diffraction analysis method on powder of BP. Prodigious intentions are quiet progressing to show the adequate potential of BP. A number of researchers have also been studied on crystal structure of BP and their transformation conditions between these allotropes of phosphorus [9–13]. In light of the rapid expanding research on BP is due to their structures, hybridization, doping and functionalization exhibits amazing characteristics to explore potential limits. First, the structure and hybridization of BP alone and also with metals, semi-metals, non-metals, etc. through mixing and stacking allow for extensible combining substances with correlative properties to design high-conduct nanostructures. Second, for peculiar applications, the intrinsic electronic structural arrangement of BP as it might be reformed by mixing metallic and non-metallic substances into the BP surface and crystal lattice. Third, the functionalizing through electrostatic, covalent and weak van der Waals interactions via surface alteration towards inorganic layers could and organic groups stabilize BP whilst at the same instant it maintains and even promotes the specificity of BP in major applications [14].

In recent years, layered BPs have been mechanically exfoliated into different layers as monolayer or few-layer BP (phosphorene) resulting from weak van der Waals interactions between their deformed layers [15–17]. Possessing a direct band gap tunability property, BP presents a band tunable band gap from bulk to single layer that is, it ranges from 0.3 to 1.5 eV reported by a number of researchers depending on the interlayer stacking pattern [18–22], enhanced anisotropic in-plane properties and charge carrier mobility and thereby, carrying a large photon absorption aperture up to the mid-infrared system of the solar spectrum. The band gap values indicate that BP possesses a high potential and wide range of light absorption in photo detector system [23–28]. The comparison of the size of the band gap of BP with graphene and similar materials, BP possesses better metal–semiconductor transformation, and that caused structural changes in the layer's number and their stress level which is not found in graphene and other 2D substances [16, 29]. Containing tunable/directed band gap properties, BP has broad applications in the area of catalysis.

Further, BP is currently at the forepart of research targeting applications related to energy production, conversion and storage, such as oxygen evolution reaction (OER), water splitting photocatalyst, solar cell, lithium/sodium-ion batteries, lithium-sulfur battery, and be possessed of its unique structure and excellent physical properties [30, 31]. Hence, it can be say that layered BP as a new wide area of research among all the family of 2D materials mainly in the field of energy production and storage devices.

BP nanostructures prepared from chemical and mechanical (liquid-phase) exfoliation methods demonstrate great advantage in many significant fields, including field-effect transistors, memory devices, diodes, demodulators and photodetectors [21, 32–37]. These properties reveal that BP as an advance material for research in the field with moderate band gap-based nano-electronic, nanophotonic devices,

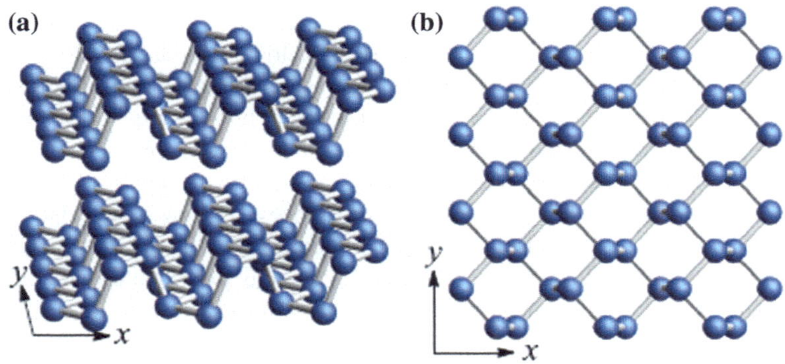

Fig. 1 **(a)** Layer structure of phosphorous. The *x* and *y* axes are perpendicular and parallel with the ridge direction, as shown. **(b)** Single layer structure of phosphorene (top view) [20]

and incorporating 2D layered materials applications. Prodigious efforts are devoted on the fundamental properties to provide a deep knowledge about this substance, [38, 39] and a number of important applications of BP including optoelectronics, transistors, sensors, thermoelectric applications are in account [9, 40–43].

It has also been seen that crystal structure of bulk BP, which is also called as phosphorene contains their layers than ten in number can be obtained from BP bulk crystal through a top-down pattern (e.g., mechanical and liquid exfoliation) [37, 44, 45]. It is until 2014, when phosphorene was successfully exfoliated from its bulk counterpart, and this 2D material generated a lot of interest in the research community [19–24]. Phosphorene contains a lot of outstanding properties due to their unique structure. In comparison with other 2D layered nanomaterials, nanoscale BP can be distinguished with their puckered structures towards the armchair direction and its double layered arrangement towards the zigzag direction as shown in Fig. 1 [13]. This structural anisotropy leads to some specific features of BP, such as thermoelectric and topological features, electronic conductivity and unusual mechanical behavior (negative Poisson's ratio) [46–51]. Among those 2D materials, graphene is the most shining material, Geim et al., first obtained from a plague of graphite in 2004 [52]. Although, graphene possesses the greater charge carrier mobility [53], it fails to play as a semiconductor because of their low band gap in its electronic structural arrangement. Another 2D material family, the transition metal dichalcogenides (TMDs) also attracted the attention of researchers which consists of a typically monolayer surface of metal atoms organized by two planes of chalcogens. After graphene, the most typical synthesized original material was Molybdenum disulfide (MoS_2), in 2008 [54], also possesses a large bandgap and resulting in large on/off ratios ($>10^8$), which allows the substance to convert electrons into quanta, or vice versa [55]. However, compared to graphene, the sandwich like structure of MoS_2 offering its mobility of charge carrier [56] and facilitates a relatively slow photo-response.

In addition to the wide bandgap of transition metals (Mo and W) based chemical substances acutely decrease in their potential applications when covers the visible region of electromagnetic spectrum. Therefore, scientists and researchers continue their work to find new 2D substances with a direct but low band gap. Besides graphene and 2D layered MoS_2, phosphorene might be considered the second 2D substances stable in free-standing form obtained from bulk black phosphorus. This is because such type of 2D hetero layered structure materials not only incorporate their peculiar properties, but also show the characteristic electronic and optical property, which are not found in isolated single layer structural substances. In view of this, large effort have been apply to understand the peculiar and eminent characteristics of 2D heterostructures that could be significantly used in nanoelectronics and optoelectronic devices, such as good efficient transistors and light-emitting diodes (LED) [57–59], and atomically thin p-n junction [9, 60–64], along with long-lived indirect physiological features [65, 66]. In addition to 2D nanosheet structure of phosphorene, BP exhibited by characteristics bond lengths and angles, large extinction coefficient in the near infrared region. In general, based on recent theoretical and experimental studies reported, these studies reveal the internal layers of BP show anisotropy in terms of stress-strain, thermal conductivity, carrier transport, phonon transmission and light absorption making BP has a number of applications [67–69]. Most recently, the other BP nanostructure, black phosphorus quantum dots (BPQDs) also come into the study to explore the significant applications of BP auction the desirable biocompatibility and photothermal properties, allows to show wide potential applications in treatment of cancer using photo-thermal technique [70–74]. In this chapter, we have reported the characteristic, structure and fundamental properties and some synthesis methods of BP for future aspects.

2 Structure and Fundamental Properties

The first successful synthesis of the most stable phosphorus allotrope known as BP could be traced back to 1914, which came from RP, when applied to high pressure and high temperature [7]. BP shows a crystal structure which was first investigated in 1935 by Hultgren et al. from the X-ray diffraction studies of BP powder [8]. BP is shiny black and thermodynamically stable allotrope. Similar to graphite, it possesses good conduction of heat and flow of electrons but shows little diamagnetic behavior as compared to RP and WP [5]. So, recent studies on the structure and fundamental properties of black phosphorus are summarized in the following section.

2.1 Structural Characteristics and Hybridization

In a unit cell, BP contains an orthorhombic structure [75] with eight phosphorus atoms. Monolayer BP contains a single layer of two atoms with a puckered honeycomb lattice structure as shown in Fig. 2 as similar to graphite which also consists of layered structure. This structure confirms that BP is the most stable allotrope than to

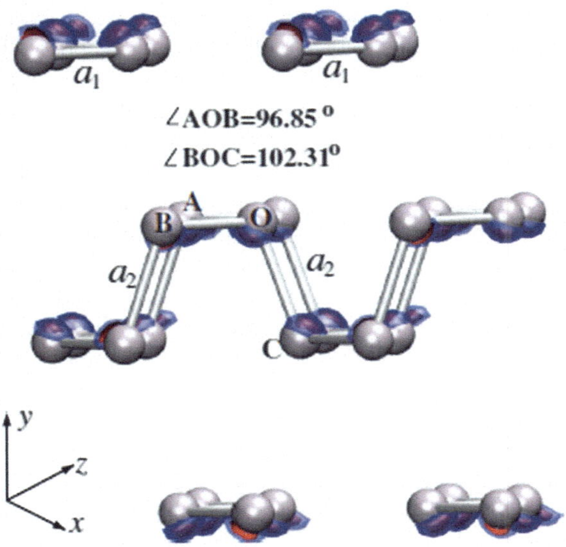

Fig. 2 Constant-density surface plot of BP in which spherical balls are considered as phosphorus atoms at atomic level while two charge density iso-surfaces with different density values are closely attached to each P atom [76]

WP and RP under normal conditions. The spatial positions of each P atom in a unit cell are denoted as:

$$uv_0 = \frac{1}{2} + u = \frac{1}{2} - v,\ 0;\ u,\ \frac{1}{2} + v,\ \frac{1}{2};\ \frac{1}{2} - u,\ v,\ \frac{1}{2}$$
$$\bar{u}\bar{v}_0 = \frac{1}{2} - u;\ \frac{1}{2} + v,\ 0;\ \bar{u},\ \frac{1}{2} + v,\ \frac{1}{2};\ \frac{1}{2} + u,\ \bar{v},\ \frac{1}{2}$$

The measurable quantities of a crystal lattice for a unit cell are listed in Table 1 [76]. Considering the structure of phosphorus, each P atoms are bonded with the neighboring three P atoms along with an unpaired electron which makes it reactive.

Table 1 Structural parameters of BP normal condition [76]

	Experimental	Theoretical
a (Å)	4.376	4.422
b (Å)	10.478	10.587
c (Å)	3.314	3.348
U	0.0806	0.0821
V	0.1017	0.1011
Bond angle and length (Å)	96.34°	96.85°
	102.09°	102.31°
	2.224	a_1 = 2.238
	2.244	a_2 = 2.261

The interlayer is bonded by weak van der Waals forces of attraction whereas the in-plane P atoms are bonded with strong covalent bond. Figure 2 and Table 1 represent that bond length, a_1 between each P–P as shown in XZ-plane is shorter than a_2 that along the Y-axis. Similarly, the bond angle is also shorter (96.85°) along the XZ-plane than that of Y-axis (102.31°). The P–P bond length was found to be 3.62 Å between the layers of BP, which is larger than that of each P–P covalent bond which was around 2.2 Å. This confirms that due to larger bond length, the bond attractions are not exist between layers and only the van der Waal's interactions make the layers are packed to one another [47, 76, 77]. This is the reason that BP atoms are formed layers structure as similar to graphite due to strong interactions in-plane. These weak van der Waals interactions which are differ from ionic and covalent interactions among BP layers which are caused by polarizations of nearby atoms. Unlike to graphite, (in which each C-atom is bonded to three neighboring C-atoms in a layer and shows sp^2 hybridization), in BP, the valence shell electronic configuration of phosphorus is $3\ s^2 3p^3$. Phosphorus has five electrons available for bonding. Each phosphorus atom is tetrahedrally bonded with neighbouring three P-atoms and shows sp^3 hybridization. This makes the P atom are arranged like as to form a puckered honeycomb structure. Each phosphorus atom has unshared pair of electrons ($3s^2$, ground state configuration) that make phosphorus very reactive towards air. The crystal structure of black phosphorus is shown in Fig. 2. The arrangement of atoms within the crystal lattice of black phosphorus yields two inequivalent directions: the zigzag (ZZ, parallel to the atomic ridges) and the armchair (AC, perpendicular to the ridges) [46, 76]. This strong structural anisotropy explains the enormous fundamental properties which will better explain in later perspective.

2.2 *Fundamental Properties*

2.2.1 Mechanical Properties

Various investigation on mechanical properties reveal that under strain, BP contain excitable electronic arrangement, of high mechanical ductility, and extremely anisotropic in nature. It was found that on the structural analysis, BP possesses good ductile strain in the armchair (30%) and also in zigzag (27%) directions [5]. The mechanical anisotropy of BP nanosheets contains superior mechanical flexibility in Young's modulus (44 GPa in armchair and 166 GPa in zigzag directions) compared with other 2D materials like graphene (1000 GPa) [29, 78]. Furthermore, the Young's modulus values reveal that monolayer BP is more ductile in x-direction than that of y-direction because of large mechanical strength as shown in Fig. 3 [79]. Compared with other 2D materials, the excellent flexibility of single-layer BP is due to the smaller Young's modulus [80]. It is noticed that by applying the tensile strain to inflect the electronic properties of phosphorene, while recent study on mechanical anisotropy revealed that when axial strain is applied appropriately, the layered structure of BP can regular show a transition which make the band gap from direct to

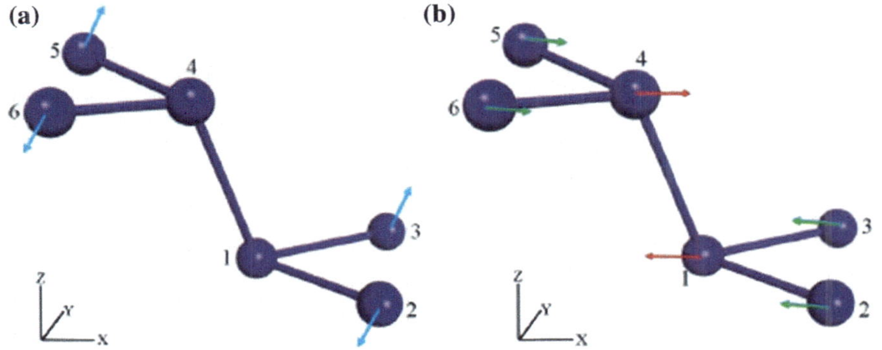

Fig. 3 **(a)** Blue arrow indicated that BP is stretched towards Y-direction, i.e., the atoms are moving towards the attached arrows. **(b)** To accommodate the tension in the y-direction, red arrow showed that BP contracts in the *x*-direction, i.e., atoms 1 and 4 move inward along the attached arrows [29]

indirect [81] and also change the order of transportation of charge [82]. Peng et al. have been studied on the tensile strength on layered structure on phosphorene and found that the 2D BP can maintain a ductile strain up to 30%, which is better than the ductile strain limits of other 2D materials such as graphene and molybdenum disulfide [50, 78]. Since, due to the honeycomb structure of phosphorene, it is more ductile in AC direction and auxetic materials have been found rather unusual the negative Poisson's ratio value [29]. Besides Young's modulus, the Poisson's ratio also explains the mechanical anisotropy in phosphorene [78]. Further, it can be seen that the tensile strain of phosphorene with other 2D materials, MoS_2 and graphene show their ductility more than 25% than phosphorene while Young's modulus values of phosphorene is lower than those of MoS_2 and graphene is due to the weak interactions P-P bond. The extension in bond angles and as compared to linear bond with layer bond that effects on tensile stress applications of phosphorene, i.e., the tensile strain extends the puckered structure of phosphorene thereby reduced the strain energy. The mechanical anisotropy of BP is still extant and has been applied in the nanoelectromechanical devices and the device shows excellent nanoelectromechanical attributes [83].

2.2.2 Electrical Properties

Both the form of BP, single as well as multilayer form makes it an ideal semiconductor device for various potential applications as in efficient photo-electrical conversion and extraordinary light emission due to its direct but tunable bandgap properties. The layer number denotes the band gap strength of BP whether it is strong or not [84, 85] due to the layer-layer coupling effect. The value of intrinsic band gap is about 0.32–0.36 eV in the Bulk BP [81, 86, 87]. A number of theoretical determinations confirm that for mono- or multilayer BP, the band gap was found to be as much as

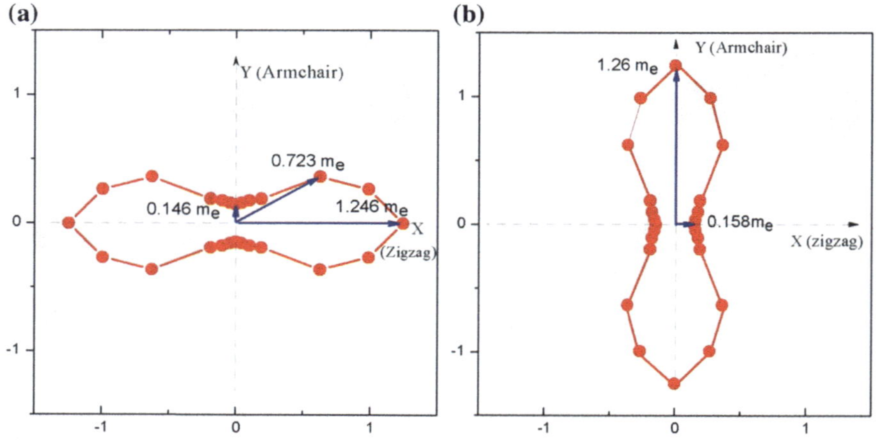

Fig. 4 Electron effective mass of 2D BP is shown into spatial direction. **(a)** Intrinsic phosphorene and **(b)** 5% biaxially strained phosphorene. Blue line arrow represents the absolute value of effective mass [82]

1.5–2.0 eV [19, 20, 88–92], which is characterized to the interactions inside the layers with an increase in the number of layers. These values show that mono/multilayer BP have high value of bandgap as compared to its bulk form. Possesses orthorhombic puckered structure, the effective mass of carriers (electrons) in ZZ direction is 10 times more than that in AM direction in BP [93]. The effective mass of electrons show good anisotropy and look like as numeral "8" (Fig. 4a) in real space for intrinsic single layer phosphorene, [82] which were predicted by Fei et al. These effective mass values can differ along different directions by an order of magnitude attributed to the anisotropic dispersion of BP. The anisotropic electric conductance is observed due to the change in effective mass or band dispersion in phosphorene [94]. The applied stress fields in both ZZ and AC directions changes the carrier mobility and energy band structure of BP. Keeping the same shape, it can be observed from Fig. 4b, while applying an appropriate uni- or biaxial strain (e.g., 5% biaxial strain), the spatial anisotropy of the effective mass of carriers and the direction of mobility of electrons is rotated by right angle (90°). The electrons contain large effective mass towards the AC direction as compared to the ZZ direction, respectively whereas the anisotropy of holes still undisturbed at all during this strains process.

2.2.3 Thermal Conductivity Properties

Phonons chiefly covered the thermal conductivity of bulk black phosphorus as comparatively low electrical conductivity of electrons [95]. Thermal conductivity anisotropy of bulk BP is also high as similar to electrical anisotropy. These thermal anisotropic mainly depends on the size, temperature and also strain of the 2D materials. Zhu et al. [96] reported the size effect of bulk phosphorus on the thermal conductivity (G) value in their study. They were found along the ZZ direction, the G value was size dependent whereas size-independent along the AC direction. BP exhibits strong G value along the AC direction than to ZZ direction by neutron diffraction and high temperature X-ray crystal diffraction were reported by Reidner in 1974. Tensile strain also plays a vital role to reduce the G values by a substantial factor in AC and ZZ directions [97]. Jain et al. studied on phosphorene and reported the G value decreases by a factor of ~4.7 when uniaxial tensile strain (~6%) is applied towards the AC direction, whereas same stain reduced the G value by ~2.2 towards ZZ direction [98]. The temperature dependent G values of phosphorene are normally depends on phonons. In accordance to power law, the thermal conductivity increases as decrease in temperature or vice versa [99]. The thermal conductivity is also measured by Yang et al. [100]. They were correlated the orthogonal relation of the extended thermal and electrical conduction in bulk BP, i.e., the ratio of electrical to thermal conductance that enhance the conductivity properties, which may be used as in thermoelectricity. 2D BP, thus contains a higher thermoelectric merit (ZT) figure in AC direction. Further study also reveals that the ZZ direction in 2D BP possesses high thermal conductivity value whereas the AC direction exhibits the high electrical conductivity value [101]. On Comparison of graphene and molybdenum disulfide with 2D BP, the G value of 2D BP is equivalent to MoS_2 [102] but lower than that of graphene [103].

The small electrical resistivity and low electrical conductivity can be characterized on the basis of seedbeck coefficient as the values of this coefficient for 2D BP are lying between low value for graphene and high for MoS_2 suggests the 2D BP has larger thermoelectric merit (ZT) figure and good thermoelectric substance. The electrical resistivity of 2D BP may be reduced substantially using electrostatic field effect and hence it is used for thermoelectric applications [46].

2.2.4 Optical Properties

Under anisotropic optical absorption, the basic optical properties of a substance mainly their narrow/directed band gap, photoconductivity, excitonic effect is generally characterized by their linear absorption measurement while under tunable optical property, highly desired characteristic for semiconductors and their photonic devices has come into the consideration. A number of parameters, such as dimension of material, doping, and strain, are considered in the modification of the anisotropic optical spectra of quasi-2D BP. BP possesses anisotropic optical property in which in their AC direction, the polarization of light is more absorbed than to ZZ direction as towards the zigzag direction, the polarization of light is transparent [20, 90]. Since, absorbing infrared as well as a little amount of UV-visible light, BP can be used as linear optical polarizer [103]. With respect to graphene and molybdenum disulfide, phosphorene, the bulk 2D BP, the time to respond and also the photo-response are lying in between the two because of their bridging gap are between the two [46]. The absorption spectra of optical anisotropy of bulk BP have been investigated by Tran

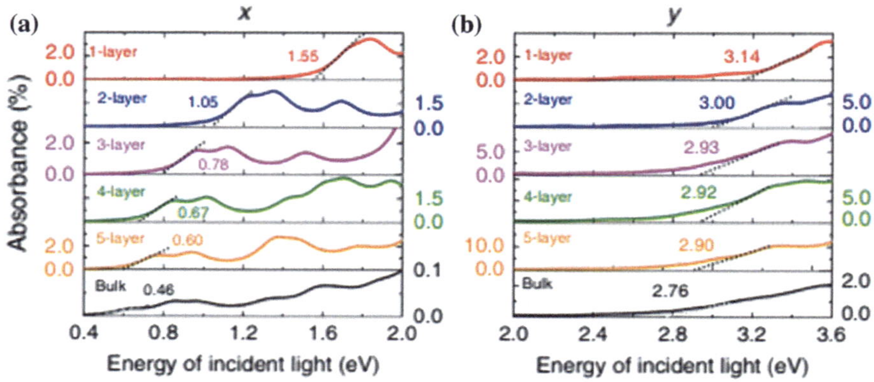

Fig. 5 **a**, **b** Optical absorption spectra of few-layer phosphorene for light polarized along the armchair direction (**a**) and zigzag direction (**b**). Black dashed lines represent an approximate linear fit to estimate the band edges for the (highly anisotropic) first absorption peaks [90]

et al. [20] and Qiao et al. [90]. They have found that the optical anisotropy properties of phosphorene are highly in nature than other 2D substances (Fig. 5a, b).

The theoretical predictions regarding the anisotropy nature of phosphorene was also confirmed by Qiao et al. [90] and Tran et al. [20] under photoluminescence measurements on BP's direct band gap that depends on the number of layer. They have found that the absorption of light polarized in AC direction has strong extinction ability than ZZ direction. It was also proposed that the light absorption phenomenon can also help to analyze the layered structure and orientation of phosphorene experimentally. These measurements found that distinct peak only in the AC direction at about the band gap ~0.3 eV, which was not observed in ZZ direction [20]. The polarized optical spectroscopy method on microscopic level used by Mao et al. [104] and the groups have been first time studied on the characteristics optical anisotropy property in the visible region for BP. Lan et al. [105] experimentally investigated the same property of BP using the conventional measurement spectroscopy and interferometric spectroscopy/imaging technique. This optical anisotropy provides the identification of surface characterization as well as lattice nature of the layered BP crystal and also applied in the area of polarization-dependent linear and non-linear optics [104].

It is to summarized that that all the studied anisotropy under structural and fundamental properties finally revealed about the puckered honeycomb structure of BP, resulting the tetrahedrally arrangement containing sp^3 nonequivalent hybridization of each P atom. These lead to the remarkable mechanical, thermal, electrical and optical anisotropic properties of single- and multilayered 2D BP [16, 20, 29, 82, 90]. The study on the anisotropic properties of BP gives an approach to better understand for novel BP-based opto-electrical devices. Beside these, optical anisotropy highly recommends BP towards optoelectronic functions because these properties supply the other degree of freedom for scientific communities to design the more efficient opto-electrical and electronic systems.

3 Synthesis of Black Phosphorus

A brief study about the synthesis of black phosphorus have been made by us. Black phosphorus is considered thermodynamically more stable allotropes of phosphorus among all under standard conditions. This is the reason that it synthesized more difficulty as compared to WP and RP [3, 4]. BP was first modified by Bridgeman (1914) from white phosphorus under temperature on/above 200 °C and pressure around 1.2 GPa at about 5–30 min [7]. In 1950, another study made by Keyes et al., [95] who tried to synthesized the BP from WP at the same temperature but increased the pressure, i.e., 1.3 GPa and found that polycrystalline BP while in case of Bridgeman, BP was in amorphous form. BP can also prepare from RP (Fig. 6) [106]. For this, first of all to remove the contaminants and surface oxides, RP was pretreated at 200 °C for 2 h. After that, the RP was obtained in the powder form which on vapourization on 450 °C for few hours. When temperature up to or above 650 °C and at high pressure level of 8.0 GPa, BP was obtained from RP. A lot of study has been focused by science community on the preparation and synthesis of BP either from WP or from RP. Krebs and Weitz synthesized BP using mercury or bismuth flux method and WP or RP was converted to BP using mercury as a catalyst [107]. Maruyama et al. [108] have taken a reaction kettle and dissolved WP in solutions contains bismuth into it and placed at 400 °C temperature for 20 h. After slow cooling. a rod-shaped BP crystal were obtained with dimension of 5 mm × 0.1 mm × 0.07 mm (grain size). Narita et al. [109] also studied on the conversion of RP to BP. They have taken pressure under 1.0 GPa, melting of RP was found at 900°C and when temperature decreased to 600 °C,

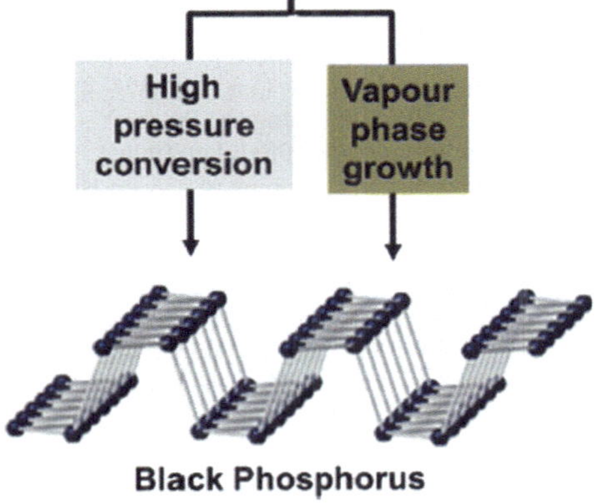

Fig. 6 Crystal structure of RP and schematic representation for the synthesis of black phosphorus from red phosphorus [106]

it was slowly cooled at a low rate around 0.5 °C/min, the BP crystal was obtained with a dimension of grain size is about 0.5 cm × 0.5 cm × 1 cm. With RP, SnI_4 and AuSn, Lange et al. also prepared BP crystals [110] under a low-pressure and high temperature conditions; however, small amount of by-products (RP and Au_3SnP_4) also obtained along with BP. Further, a detail study in synthesis strategy of BP are given in the next chapter.

4 Conclusion

In the present chapter, we have discussed the structural and fundamental properties and a brief idea about the synthesis of black phosphorus and compared their structure with other 2D materials. We have also discussed the attractive properties of single or few layered BP which make it unique and compare with the other 2D materials such as graphene and TMDs. We have seen that BP contains relatively thermodynamic stability and presents a band tunable band gap depending on the interlayer stacking pattern in which the band gap may be balanced by layer's number and stress. BP also contains various anisotropy property, for example, mechanical, electrical, thermal conductivity and optical anisotropies, which confirms BP widely used in optoelectronics, transistors, sensors, thermoelectrical devices and many other fields. Under structure and hybridization, the three valence electrons ($2p^3$) in each P-atom makes it behave like a semiconductor. BP shows semiconducting property due to direct band gap and also tunable as compare to graphene, as a zero band gap semiconductor, according to the applied strain. In addition to 2D nanosheet structure of phosphorene, BP exhibited by different bond lengths, strengths and angles, large extinction coefficient in the near IR region.

References

1. Sun, J., Zheng, G., Lee, H.W., Liu, N., Wang, H., Yao, H.: Formation of stable phosphorus-carbon bond for enhanced performance in black phosphorus nanoparticle-graphite composite battery anodes. Nano Lett. **14**, 4573–4580 (2014). https://doi.org/10.1021/nl501617j
2. Sun, C., Wen, L., Zeng, J., Wang, Y., Sun, Q., Deng, L., Zhao, C., Li, Z.: One-pot solvent less preparation of PEGylated black phosphorus nanoparticles for photoacoustic imaging and photothermal therapy of cancer. Biomaterials **91**, 81–89 (2016). https://doi.org/10.1016/j.biomaterials.2016.03.022
3. Gui, R., Jin, H., Wang, Z., Li, J.: Black phosphorus quantum dots: synthesis, properties, functionalized modification and applications. Chem. Soc. Rev. **47**, 6795–6823 (2018). https://doi.org/10.1039/C8CS00387D
4. Choi, W., Lahiri, I., Seelaboyina, R., Kang, Y.S.: Synthesis of graphene and its applications: A review. Crit. Rev. Solid State Mater. Sci. **35**, 52–71 (2010). https://doi.org/10.1080/10408430903505036
5. Khandelwal, A., Mani, K., Karigerasi, M.H., Lahiri, I.: Phosphorene-the two-dimensional black phosphorous: properties, synthesis and applications. Mater. Sci. Eng. B. **221**, 17–34 (2017). https://doi.org/10.1016/j.mseb.2017.03.011

6. Pumera, M.: Phosphorene and black phosphorus for sensing and biosensing. TrAC, Trends Anal. Chem. **93**, 1–6 (2017). https://doi.org/10.1016/j.trac.2017.05.002
7. Bridgman, P.W.: Two new modifications of phosphorus. J. Am. Chem. Soc. **36**, 1344–1363 (1914). https://doi.org/10.1021/ja02184a002
8. Hultgren, R., Gingrich, N.S., Warren, B.E.: The atomic distribution in red and black phosphorus and the crystal structure of black phosphorus. J. Chem. Phys. **3**, 351 (1935). https://doi.org/10.1063/1.1749671
9. Li, L.K., Yu, Y.J., Ye, G.J., Ge, Q.Q., Ou, X.D., Wu, H., Feng, D.L., Chen, X.H., Zhang, Y.B.: Black phosphorus field-effect transistors. Nat. Nanotechnol. **9**, 372–377 (2014). https://doi.org/10.1038/nnano.2014.35
10. Late, D.J.: Liquid exfoliation of black phosphorus nanosheets and its application as humidity sensor. Microporous Mesoporous Mater. **225**, 494–503 (2016). https://doi.org/10.1016/j.micromeso.2016.01.031
11. Zhao, W., Xue, Z., Wang, J., Jiang, J., Zhao, X., Mu, T.: Large-Scale, Highly efficient, and green liquid-exfoliation of black phosphorus in ionic liquids. ACS Appl. Mater. Interfaces **7**, 27608–27612 (2015). https://doi.org/10.1021/acsami.5b10734
12. Xu, J.Y., Gao, L.F.C., Hu, X., Zhu, Z.Y., Zhao, M., Wang, Q., Zhang, H.L.: Preparation of large size, few-layer black phosphorus nanosheets via phytic acid-assisted liquid exfoliation. Chem. Commun. **52**, 8107–8110 (2016). https://doi.org/10.1039/C6CC03206K
13. Zhang, R., Zhou, X.Y., Zhang, D., Lou, W.K., Zhai, F., Chang, K.: Electronic and magneto-optical properties of monolayer phosphorene quantum dots. 2D Mater. **2**, 045012 (2015). https://doi.org/10.1088/2053-1583/2/4/045012/meta
14. Lei, W., Liu, G., Zhang, J., Liu, M.: Black phosphorus nanostructures: recent advances in hybridization, doping and functionalization. Chem. Soc. Rev. **46**, 3492–3509 (2017). https://doi.org/10.1039/C7CS00021A
15. Appalakondaiah, S., Vaitheeswaran, G., Lebegue, S., Christensen, N.E., Svane, A.: Effect of van der Waals interactions on the structural and elastic properties of black phosphorus. Phys. Rev. B: Condens. Matter Mater. Phys. **86**, 035105 (2012). https://doi.org/10.1103/physrevb.86.035105
16. Xia, F., Wang, H., Jia, Y.: Rediscovering black phosphorus as an anisotropic layered material for optoelectronics and electronics. Nat. Commun. **5**, 4458 (2014). https://doi.org/10.1038/ncomms5458
17. Cakir, D., Sahin, H., Peeters, F.M.: Tuning of the electronic and optical properties of single-layer black phosphorus by strain. Phys. Rev. B. **90**, 205421 (2014). https://doi.org/10.1103/PhysRevB.90.205421
18. Sun, Z., Martinez, A., Wang, F.: Optical modulators with two-dimensional layered materials. Nat. Photon. **10**, 227–238 (2016). https://doi.org/10.1038/NPHOTON.2016.15
19. Low, T., Rodin, A.S., Carvalho, A., Jiang, Y., Wang, H., Xia, F., Neto, A.H.C.: Tunable optical properties of multilayer black phosphorus thin films. Phys. Rev. B. **90**, 075434–1–075434–5 (2014). https://doi.org/10.1103/physrevb.90.075434
20. Tran, V., Soklaski, R., Liang, Y., Yang, L.: Layer-controlled band gap and anisotropic excitons in few-layer black phosphorus. Phys. Rev. B. **89**, 817–824 (2014). https://doi.org/10.1103/PhysRevB.89.235319
21. Chen, Y., Ren, R., Pu, H., Chang, J., Mao, S., Chen, J.: Field-effect transistor biosensors with two-dimensional black phosphorus nanosheets. Biosens. Bioelectron. **89**, 505–510 (2016). https://doi.org/10.1016/j.bios.2016.03.059
22. Li, D., Castillo, A.E.D.R., Jussila, H., Ye, G., Ren, Z., Bai, J., Chen, H.L., Sun, Z., Bonaccorso, F.: Black phosphorus polycarbonate polymer composite for pulsed fibre lasers. Appl. Mater. Today **4**, 17–23 (2016). https://doi.org/10.1016/j.apmt.2016.05.001
23. Tan, W.C., Huang, L., Ng, R.J., Wang, L., Hasan, D.M.N., Duffin, T.J., Kumar, K.S., Nijhuis, C.A., Lee, C., Ang, K.W.: A black phosphorus carbide infrared phototransistor. Adv. Mater. **30**, 1705039 (2018). https://doi.org/10.1002/adma.201705039
24. Bullock, J., Amani, M., Cho, J., Chen, Y.Z., Ahn, G.H., Adinolfi, V., Shrestha, V.R., Gao, Y., Crozier, K.B., Chueh, Y.L., Javey, A.: Polarization-resolved black phosphorus/molybdenum

disulfide mid-wave infrared photodiodes with high detectivity at room temperature. Nat. Photon. **12**, 601–607 (2018). https://doi.org/10.1038/s41566-018-0239-8
25. Miao, J., Song, B., Xu, Z., Cai, L., Zhang, S., Dong, L., Wang, C.: Single pixel black phosphorus photodetector for near-infrared imaging. Small **14**, 1702082 (2018). https://doi.org/10.1002/smll.201702082
26. Huang, Y., Liu, X., Liu, Y., Shao, Y., Zhang, S., Fang, C., Han, G., Zhang, J., Hao, Y.: Nanostructured multiple-layer black phosphorus photodetector based on localized surface plasmon resonance. Optic. Mater. Express **9**, 739–750 (2019). https://doi.org/10.1364/OME.9.000739
27. Viti, L., Politano, A., Zhang, K., Vitiello, M.S.: Thermoelectric terahertz photodetectors based on selenium-doped black phosphorus flakes. Nanoscale **11**, 1995–2002 (2019). https://doi.org/10.1039/c8nr09060b
28. Dhanabalan, S.C., Ponraj, J.S., Guo, Z., Li, S., Bao, Q., Zhang, H.: Emerging trends in phosphorene fabrication towards next generation devices. Adv. Sci. **4**, 1600305 (2017). https://doi.org/10.1002/advs.201600305
29. Jiang, J.W., Park, H.S.: Negative poisson's ratio in single-layer black phosphorus. Nat. Commun. **5**, 4727–4734 (2014). https://doi.org/10.1038/ncomms5727
30. Fu, Y., Wei, Q., Zhang, G., Sun, S.: Advanced phosphorus-based materials for lithium/sodium-ion batteries: recent developments and future perspectives. Adv. Energy Mater. **8**, 1702849–1702867 (2018). https://doi.org/10.1002/aenm.201702849
31. Li, L., Chen, L., Mukherjee, S., Gao, J., Sun, H., Liu, Z., Ma, X., Gupta, T., Singh, C.V., Ren, W., Cheng, H.M., Koratkar, N.: Phosphorene as a polysulfide immobilizer and catalyst in high-performance lithium–sulfur batteries. Adv. Mater. **29**, 1602734–1602742 (2017). https://doi.org/10.1002/adma.201602734
32. Peng, J., Lai, Y.Q., Chen, Y.Y., Xu, J., Sun, L.P., Weng, J.: Sensitive detection of carcinoembryonic antigen using stability-limited few-layer black phosphorus as an electron donor and a reservoir. Small **13**, 1603589 (2017). https://doi.org/10.1002/smll.201603589
33. Jia, Z.Y., Xiang, J.Y., Mu, C.P., Wen, F.S., Yang, R.L., Hao, C.X., Liu, Z.Y.: J. Mater. Sci. **52**, 11506–11512 (2017)
34. Yang, Y., Gao, J., Zhang, Z., Xiao, S., Xie, H.H., Sun, Z.B., Wang, J.H., Zhou, C.H., Wang, Y.W., Guo, X.Y.: Black phosphorus-based photocathodes in wideband bifacial dye-sensitized solar cells. Adv. Mater. **28**, 8937–8944 (2016). https://doi.org/10.1002/adma.201602382
35. Rahman, M.Z., Kwong, C.W., Davey, K., Qiao, S.Z.: 2D phosphorene as a water splitting photocatalyst: fundamentals to applications. Energy Environ. Sci. **9**, 709–728 (2016). https://doi.org/10.1039/C5EE03732H
36. Bai, L., Sun, L., Wang, Y., Liu, Z., Gao, Q., Xiang, H., Xie, H., Zhao, Y.: Solution-processed black phosphorus/PCBM hybrid heterojunctions for solar cells. J. Mater. Chem. A **5**, 8280–8286 (2017). https://doi.org/10.1039/C6TA08140A
37. Zhu, M., Cai, X., Fujitsuka, M., Zhang, J., Majima, T.: Au/$La_2Ti_2O_7$ nanostructures sensitized with black phosphorus for plasmon-enhanced photocatalytic hydrogen production in visible and near-infrared light. Angew. Chem. Int. Ed. **56**, 2064–2068 (2017). https://doi.org/10.1002/anie.201612315
38. He, J., He, D., Wang, Y., Cui, Q., Bellus, M.Z., Chiu, H.Y., Zhao, H.: Exceptional and anisotropic transport properties of photocarriers in black phosphorus. ACS Nano **9**, 6436 (2015). https://doi.org/10.1021/acsnano.5b02104
39. Guo, H., Lu, N., Dai, J., Wu, X., Zeng, X.C.: Phosphorene nanoribbons, phosphorus nanotubes, and van der waals multilayers. J. Phys. Chem. C **118**, 14051 (2014). https://doi.org/10.1021/jp505257g
40. Sun, J., Lee, H.W., Pasta, M., Yuan, H., Zheng, G., Sun, Y., Li, Y., Cui, Y.: A phosphorene-graphene hybrid material as a high-capacity anode for sodium-ion batteries. Nat. Nanotechnol. **10**, 980 (2015). https://doi.org/10.1038/nnano.2015
41. Martinez, C.C.M., Sofer, Z., Pumera, M.: Layered black phosphorus as a selective vapor sensor. Angew. Chem. Int. Ed. **54**, 14317 (2015). https://doi.org/10.1002/anie.201505015

42. Wang, H., Yang, X., Shao, W., Chen, S., Xie, J., Zhang, X., Wang, J., Xie, Y.: Ultrathin black phosphorus nanosheets for efficient singlet oxygen generation. J. Am. Chem. Soc. **137**, 11376 (2015). https://doi.org/10.1021/jacs.5b06025
43. Gomez, A.C.: Black phosphorus: narrow gap, wide applications. J. Phys. Chem. Lett. **6**, 4280 (2015). https://doi.org/10.1021/acs.jpclett.5b01686
44. Liu, H., Du, Y.C., Deng, Y.X., Ye, P.D.: Semiconducting black phosphorus: synthesis, transport properties and electronic applications. Chem. Soc. Rev. **44**, 2732–2743 (2015). https://doi.org/10.1039/c4cs00257a
45. Eswaraiah, V., Zeng, Q.S., Long, Y., Liu, Z.: Black phosphorus nanosheets: synthesis, characterization and applications. Small **12**, 3480–3502 (2016). https://doi.org/10.1002/smll.201600032
46. Choi, S.J., Kim, B.K., Lee, T.H., Kim, Y.H., Li, Z., Pop, E., Kim, J.J., Song, J.H., Bae, M.H.: Electrical and thermoelectric transport by variable range hopping in thin black phosphorus devices. Nano Lett. **16**, 3969–3975 (2016). https://doi.org/10.1021/acs.nanolett.5b04957
47. Viti, L., Politano, A., Vitiello, M.S.: Thermoelectric terahertz photodetectors based on selenium-doped black phosphorus flakes. Appl. Mater. **5**, 035602 (2017). https://doi.org/10.1039/c8nr09060b
48. Tran, V., Fei, R., Yang, L.: Quasiparticle energies, excitons, and optical spectra of few-layer black phosphorus. 2D Mater. **2**, 044014 (2015). https://doi.org/10.1088/2053-1583/2/4/044014/pdf
49. Cupo, A., Meunier, V.: Quantum confinement in black phosphorus-based nanostructures. J. Phys. Condens. Matter **29**, 283001 (2017). https://doi.org/10.1088/1361-648x/aa748c
50. Wei, Q., Peng, X.: Superior mechanical flexibility of phosphorene and few-layer black phosphorus Appl. Phys. Lett. **104**, 251915 (2014). https://doi.org/10.1063/1.4885215
51. Viti, L., Hu, J., Coquillat, D., Knap, W., Tredicucci, A., Politano, A., Vitiello, M.S.: Black phosphorus terahertz photodetectors. Adv. Mater. **27**, 5567–5572 (2015). https://doi.org/10.1002/adma.201502052
52. Novoselov, K.S., Geim, A.K., Morozov, S.V., Jiang, D., Zhang, Y., Dubonos, S.V., Grigorieva, I.V., Firsov, A.A.: Electric field effect in atomically thin carbon films. Science **306**, 666–669 (2004). https://doi.org/10.1126/science.1102896
53. Chen, J.H., Jang, C., Xiao, S., Ishigami, M., Fuhrer, M.S.: Intrinsic and extrinsic performance limits of graphene devices on SiO_2. Nat. Nanotechnol. **3**, 206–209 (2008). https://doi.org/10.1038/nnano.2008.58
54. Yong, K.S., Otalvaro, D.M., Duchemin, I., Saeys, M., Joachim, C.: Calculation of the conductance of a finite atomic line of sulfur vacancies created on a molybdenum disulfide surface. Phys. Rev. B: Condens. Matter **77**, 998–1002 (2008). https://doi.org/10.1103/PhysRevB.77.205429
55. Radisavljevic, B., Radenovic, A., Brivio, J., Giacometti, V., Kis, A.: Single-layer MoS_2 transistors. Nat. Nano **6**, 147–150 (2011). https://doi.org/10.1038/nnano.2010.279
56. Fuhrer, M.S., James, H.: Measurement of mobility in dual-gated MoS_2 transistors. Nat. Nanotechnol. **8**, 146–147 (2013). https://doi.org/10.1038/nnano.2013.30
57. Lee, C.H., Lee, G.H., van Der Zande, A.M., Chen, W., Li, Y., Han, M., Cui, X., Arefe, G., Nuckolls, C., Heinz, T.F., Guo, J., Hone, J., Kim, P.: Atomically thin p-n junctions with van der Waals heterointerfaces. Nat. Nanotechnol. **9**, 676–681 (2014). https://doi.org/10.1038/nnano.2014.150
58. Yu, J.H., Lee, H.R., Hong, S.S., Kong, D.S., Lee, H.W., Wang, H.T., Xiong, F., Wang, S., Cui, Y.: Vertical heterostructure of two-dimensional MoS_2 and WSe_2 with vertically aligned layers. Nano Lett. **15**, 1031–1035 (2015). https://doi.org/10.1021/nl503897h
59. Li, M.Y., Shi, Y.M., Cheng, C.C., Lu, L.S., Lin, Y.C., Tang, H.L., Tsai, M.L., Chu, C.W., Wei, K.H., He, J.H., Chang, W.H., Suenaga, K., Li, L.J.: Epitaxial growth of a monolayer WSe_2-MoS_2 lateral p-n junction with an atomically sharp interface. Science **349**, 524–528 (2015). https://doi.org/10.1126/science.aab409
60. Huo, N.J., Kang, J., Wei, Z.M., Li, S.S., Li, J.B., Wei, S.H.: Novel and enhanced optoelectronic performances of multilayer MoS_2–WS_2 heterostructure transistors. Adv. Funct. Mater. **24**, 7025–7031 (2014). https://doi.org/10.1002/adfm.201401504

61. Cheng, R., Li, D., Zhou, H., Wang, C., Yin, A., Jiang, S., Liu, Y., Chen, Y., Huang, Y., Duan, X.: Electroluminescence and photocurrent generation from atomically sharp WSe_2/MoS_2 heterojunction p-n diodes. Nano Lett. **14**, 5590–5597 (2014). https://doi.org/10.1021/nl502075n
62. Withers, F., Zamudio, O.D.P., Mishchenko, A., Rooney, A.P., Gholinia, A., Watanabe, K., Taniguchi, T., Haigh, S.J., Geim, A.K., Tartakovskii, A.I., Novoselov, K.S.: Light-emitting diodes by band-structure engineering in van der Waals heterostructures. Nat. Mater. **14**, 301–306 (2015). https://doi.org/10.1038/nmat4205
63. Avsar, A., Vera-Marun, I.J., Tan, J.Y., Watanabe, K., Taniguchi, T., Castro Neto, A.H., Ozyilmaz, B.: Air-stable transport in graphene-contacted, fully encapsulated ultrathin black phosphorus-based field-effect transistors. ACS Nano **9**, 4138–4145 (2015). https://doi.org/10.1021/acsnano.5b00289
64. Perello, D.J., Chae, S.H., Song, S., Lee, Y.H.: High-performance n-type black phosphorus transistors with type control via thickness and contact-metal engineering. Nat. Commun. **6**, 7809 (2015). https://doi.org/10.1038/ncomms8809
65. Ceballos, F., Bellus, M.Z., Chiu, H.Y., Zhao, H.: Ultrafast charge separation and indirect exciton formation in a MoS_2-$MoSe_2$ van der Waals heterostructure. ACS Nano **8**, 12717–12724 (2014). https://doi.org/10.1021/nn505736z
66. Rivera, P., Schaibley, J.R., Jones, A.M., Ross, J.S., Wu, S., Aivazian, G., Klement, P., Seyler, K., Clark, G., Ghimire, N.J., Yan, J., Mandrus, D.G., Yao, W., Xu, X.: Observation of long-lived interlayer excitons in monolayer $MoSe_2$-WSe_2 heterostructures. Nat. Commun. **6**, 6242 (2015). https://doi.org/10.1038/ncomms7242
67. Chen, P., Li, N., Chen, X., Ong, W.J., Zhao, X.: The rising star of 2D black phosphorus beyond graphene: synthesis, properties and electronic applications. 2D Mater. **5**, 014002 (2017). https://doi.org/10.1088/2053-1583/aa8d37
68. Liu, X., Ryder, C.R., Wells, S.A., Hersam, M.C.: Resolving the in-plane anisotropic properties of black phosphorus. Small Meth. **1**, 1700143 (2017). https://doi.org/10.1002/smtd.201700143
69. Jang, H., Wood, J.D., Ryder, C.R., Hersam, M.C., Cahill, D.G.: Anisotropic thermal conductivity of exfoliated black phosphorus. Adv. Mater. **27**, 8017–8022 (2015). https://doi.org/10.1002/adma.201503466
70. Zhang, X., Xie, H., Liu, Z., Tan, C., Luo, Z., Li, H., Lin, J., Sun, L., Chen, W., Xu, Z., Xie, L., Huang, W., Zhang, H.: Black phosphorus quantum dots. Angew. Chem. Int. Ed. Engl. **54**, 3653–3657 (2015). https://doi.org/10.1002/ange.201409400
71. Sun, Z., Xie, H., Tang, S., Yu, X.F., Guo, Z., Shao, J., Zhang, H., Huang, H., Wang, H., Chu, P.K.: Ultrasmall black phosphorus quantum dots: synthesis and use as photothermal agents. Angew. Chem. Int. Ed. Engl. **54**, 11526–11530 (2015). https://doi.org/10.1002/anie.201506154
72. Xie, H., Shao, J., Ma, Y., Wang, J., Huang, H., Yang, N., Wang, H., Ruan, C., Luo, Y., Wang, Q.Q., Chu, P.K., Yu, X.F.: Biodegradable near-infrared-photoresponsive shape memory implants based on black phosphorus nanofillers. Biomaterials **164**, 11–21 (2018). https://doi.org/10.1016/j.biomaterials.2018.02.040
73. Choi, J.R., Yong, K.W., Choi, J.Y., Nilghaz, A., Lin, Y., Xu, J., Lu, X.: Black phosphorus and its biomedical applications. Theranostics **8**, 1005–1026 (2018). https://doi.org/10.7150/thno.22573
74. Liang, X., Ye, X., Wang, C., Xing, C., Miao, Q., Xie, Z., Chen, X., Zhang, X., Zhang, H., Mei, L.: Photothermal cancer immunotherapy by erythrocyte membrane-coated black phosphorus formulation. J. Control Release **296**, 150–161 (2019). https://doi.org/10.1016/j.jconrel.2019.01.027
75. Crichton, W.A., Mezouar, M., Monaco, G., Falconi, S.: Phosphorus: new in situ powder data from large-volume apparatus. Powder Diffr. **18**, 155–158 (2003). https://doi.org/10.1154/1.1545115
76. Du, Y., Ouyang, C., Shi, S., Lei M.: Ab initio studies on atomic and electronic structures of black phosphorus. J. Appl. Phys. **107**, 093718, 1–4 (2010). https://doi.org/10.1063/1.3386509

77. Takao, Y., Akira, M.: Electronic structure of black phosphorus: tight binding approach. Physica (Amsterdam) **105**, 93–98 (1981). https://doi.org/10.1016/0378-4363(81)90222-9
78. Peng, X., Wei, Q., Copple, A.: Strain-engineered direct-indirect band gap transition and its mechanism in two-dimensional phosphorene. Phys. Rev. B **90**, 085402 (2014). https://doi.org/10.1103/PhysRevB.90.085402
79. Qin, G., Yan, Q.B., Qin, Z., Yue, S.Y., Cui, H.J., Zheng, Q.R., Su, G.: Hinge-like structure induced unusual properties of black phosphorus and new strategies to improve the thermoelectric performance. Sci. Rep. **4**, 6946 (2014). https://doi.org/10.1038/srep06946
80. Jiang, J.W., Park, H.S.: J. Phys. D: Mechanical properties of single-layer black phosphorus. Appl. Phys. **47**, 385304 (2014). https://doi.org/10.1088/0022-3727/47/38/385304/meta
81. Rodin, A., Carvalho, A., Neto, A.C.: Strain-induced gap modification in black phosphorus. Phys. Rev. Lett. **112**, 176801 (2014). https://doi.org/10.1103/PhysRevLett.112.176801
82. Fei, R., Yang, L.: Strain-engineering the anisotropic electrical conductance of few-layer black phosphorus. Nano Lett. **14**, 2884–2889 (2014). https://doi.org/10.1021/nl500935z
83. Wang, Z., Jia, H., Zheng, X., Yang, R., Wang, Z., Ye, G., Chen, X., Shan, J., Feng, P.X.L.: Black phosphorus nanoelectromechanical resonators vibrating at very high frequencies. Nanoscale **7**, 877–884 (2015). https://doi.org/10.1039/C4NR04829F
84. Xu, R., Yang, J., Myint, Y.W., Pei, J., Yan, H., Wang, F., Lu, Y.: Exciton brightening in monolayer phosphorene via dimensionality modification. Adv. Mater. **28**, 3493–3498 (2016). https://doi.org/10.1002/adma.201505998
85. Brown, A., Rundqvist, S.: Refinement of the crystal structure of black phosphorus. Acta Crystallogr. **19**, 684–685 (1965). https://doi.org/10.1107/S0365110X65004140
86. Takao, Y., Morita, A.: Electronic structure of black phosphorus in tight binding approach. J. Phys. Soc. Jpn. **50**, 3362–3369 (1981). https://doi.org/10.1143/JPSJ.50.3362
87. Keyes, R.: The electrical properties of black phosphorus. Phys. Rev. **92**, 580–584 (1953). https://doi.org/10.1103/PhysRev.92.580
88. Woomer, A.H., Farnsworth, T.W., Hu, J., Wells, R.A., Donley, C.L., Warren, S.C.: Phosphorene: synthesis, scale-up, and quantitative optical spectroscopy. ACS Nano **9**, 8869–8884 (2015). https://doi.org/10.1021/acsnano.5b02599
89. Tran, V., Soklaski, R., Liang, Y., Yang, L.: Eprint Arxiv arXiv preprintar **1402**, 4192 (2014)
90. Qiao, J., Kong, X., Hu, Z.X., Yang, F., Ji, W.: High-mobility transport anisotropy and linear dichroism in few-layer black phosphorus. Nat. Commun. **5**, 4475 (2014). https://doi.org/10.1038/ncomms5475
91. Chen, Y., Jiang, G., Chen, S., Guo, Z., Yu, X., Zhao, C., Zhang, H., Bao, Q., Wen, S., Tang, D., Fan, D.: Mechanically exfoliated black phosphorus as a new saturable absorber for both Q-switching and mode-locking laser operation. Opt. Express **23**, 12823–12833 (2015). https://doi.org/10.1364/OE.23.012823
92. Gomez, A.C., Vicarelli, L., Prada, E., Island, J.O., Narasimha-Archarya, K.L., Blanter, S.I., Groenendijk, D.J., Buscema, M., Steele, G.A., Alvarez, J.V., Zandbergen, H.W., Palacios, J.J., van der Zant, H.S.J.: Isolation and characterization of few-layer black phosphorus. 2D Mater. **1**, 025001 (2014). https://doi.org/10.1088/2053-1583/1/2/025001
93. Ling, X., Wang, H., Huang, S., Xia, F., Dresselhaus, M.S.: The renaissance of black phosphorus. Proc. Natl. Acad. Sci. USA **112**, 4523–4530 (2015). https://doi.org/10.1073/pnas.1416581112
94. Liu, H., Neal, A.T., Zhu, Z., Luo, Z., Xu, X., Tomanek, D., Ye, P.D.: Phosphorene: an unexplored 2D semiconductor with a high hole mobility. ACS Nano **8**, 4033–4041 (2014). https://doi.org/10.1021/nn501226z
95. Lv, H.Y., Lu, W.J., Shao, D.F., Sun, Y.P.: Enhanced thermoelectric performance of phosphorene by strain-induced band convergence. Phys. Rev. B **90**, 085433, 1–8 (2014). https://doi.org/10.1103/physrevb.90.085433
96. Zhu, L., Zhang, G., Li, B.: Coexistence of size-dependent and size-independent thermal conductivities in phosphorene. Phys. Rev. B **90**, 214302, 1–6 (2014). https://doi.org/10.1103/physrevb.90.214302

97. Zhang, Y.Y., Pei, Q.X., Jiang, J.W., Wei, N., Zhang, Y.W.: Thermal conductivities of single- and multi-layer phosphorene: a molecular dynamics study. Nanoscale **8**, 483–491 (2015). https://doi.org/10.1039/C5NR05451F
98. Jain, A., McGaughey, A.J.H.: Strongly anisotropic in-plane thermal transport in single-layer black phosphorene. Sci. Rep. **5**, 8501 (2015). https://doi.org/10.1038/srep08501
99. Qin, G., Yan, Q.B., Qin, Z., Yue, S.Y., Hu, M., Su, G.: Anisotropic intrinsic lattice thermal conductivity of phosphorene from first principles. Phys. Chem. Chem. Phys. **17**, 4854–4858 (2015). https://doi.org/10.1039/C4CP04858J
100. Fei, R., Faghaninia, A., Soklaski, R., Yan, J.A., Lo, C., Yang, L.: Enhanced thermoelectric efficiency via orthogonal electrical and thermal conductances in phosphorene. Nano Lett. **14**, 6393–6399 (2014). https://doi.org/10.1021/nl502865s
101. Ong, Z.Y., Cai, Y., Zhang, G., Zhang, Y.W.: Strong thermal transport anisotropy and strain modulation in single-layer phosphorene. J. Phys. Chem. C **118**, 25272–25277 (2014). https://doi.org/10.1021/jp5079357
102. Cai, Y., Lan, J., Zhang, G., Zhang, Y.W.: Lattice vibrational modes and phonon thermal conductivity of monolayer MoS_2. Phys. Rev. B **89**, 035438 (2014). https://doi.org/10.1103/PhysRevB.89.035438
103. Xu, Y., Chen, X., Gu, B.L., Duan, W.: Intrinsic anisotropy of thermal conductance in graphene nanoribbons. Appl. Phys. Lett. **95**, 233116 (2009). https://doi.org/10.1063/1.3272678
104. Mao, N., Tang, J., Xie, L., Wu, J., Han, B., Lin, J., Deng, S., Ji, W., Xu, H., Liu, K., Tong, L., Zhang, J.: Optical anisotropy of black phosphorus in the visible regime. J. Am. Chem. Soc. **138**, 300–305 (2016). https://doi.org/10.1021/jacs.5b10685
105. Lan, S., Rodrigues, S., Kang, L., Cai, W.: Visualizing optical phase anisotropy in black phosphorus. ACS Photon. **3**, 1176–1181 (2016). https://doi.org/10.1021/acsphotonics.6b00320
106. Martinez, C.C.M., Sofer, Z., Sedmidubsky, D., Luxa, J., Kherzia, B., Pumera, M.: Metallic impurities in black phosphorus nanoflakes prepared by different synthetic routes. Nanoscale **10**, 1540–1546 (2018). https://doi.org/10.1039/c7nr05718k
107. Krebs, H., Weitz, H., Worms, K.H., Anorg. Z.: About the structure and properties of semimetals. VIII. The catalytic representation of black phosphorus. Allg. Chem. **280**, 119–133 (1955). https://doi.org/10.1002/zaac.19552800110
108. Maruyama, Y., Suzuki, S., Kobayashi, K., Tanuma, S.: Synthesis and some properties of black phosphorus single crystals. Physica B + C **105**, 99–102 (1981). https://doi.org/10.1016/0378-4363(81)90223-0
109. Narita, S., Terada, S., Mori, S., Muro, K., Akahama, Y., Endo, S.: Far-Infrared cyclotron resonance absorptions in black phosphorus single crystals. J. Phys. Soc. Jpn. **52**, 3544–3553 (1983). https://doi.org/10.1143/jpsj.52.3544
110. Lange, S., Schmidt, P., Nilges, T.: Au_3SnP_7@black phosphorus: an easy access to black phosphorus. Inorg. Chem. **38**, 4028 (2007). https://doi.org/10.1021/ic062192q

Future Prospects and Challenges of Black Phosphorous Materials

Zahra Azizi, Mohammad Ghashghaee and Mehdi Ghambarian

Abstract The cutting-edge developments in the field of black phosphorus (BP) nanostructures have contributed significantly to the progress of 2D nanomaterials in a broad range of foreseeable applications. This chapter intends to outline the remaining challenges and prospects of different BP nanomaterials, including the bulk phase, few-layer BP structures, nanoribbons, nanotubes, and heterostructures. Potential perspectives in different application areas including but not limited to electronic devices, sensors, biomedical devices, and catalysis are briefly reviewed.

Abbreviations

2D	Two-dimensional
AFM	Atomic force microscopy
AIBN	Azodiisobutyronitrile
ALD	Atomic layer deposition
BP	Black phosphorus
CVD	Chemical vapor deposition
DFT	Density-functional theory
FET	Field-effect transistor
GNR	Graphene-based nanoribbon
h-BN	Hexagonal boron nitride
IR	Infrared

Z. Azizi
Department of Chemistry, Karaj Branch, Islamic Azad University, P.O. Box 31485-313, Karaj, Iran

M. Ghashghaee
Faculty of Petrochemicals, Iran Polymer and Petrochemical Institute, P.O. Box 14975-112, Tehran, Iran

M. Ghambarian (✉)
Gas Conversion Department, Faculty of Petrochemicals, Iran Polymer and Petrochemical Institute, P.O. Box 14975-112, Tehran, Iran
e-mail: m.ghambarian@ippi.ac.ir

Inamuddin et al. (eds.), *Black Phosphorus*, Engineering Materials,
https://doi.org/10.1007/978-3-030-29555-4_8

LMH	Layered metal hydroxide
MD	Molecular dynamics
MOF	Metal–organic framework
NMP	N-methylpyrrolidone
OFET	Organic field effect transistor
OLED	Organic light emitting diodes
OPV	Organic photovoltaic materials
PDDA	Poly dimethyldiallyl ammonium chloride
PNR	Phosphorene nanoribbon
RP	Red phosphorus
SAC	Single atom catalyst
STEM	Scanning transmission electron microscopy
STM	Scanning tunnelling microscopy
TMD	Transition metal dichalcogenide
vdW	van der Waals

1 Introduction

Among the two-dimensional (2D) materials from group VA family, the most investigated member is black phosphorus (BP), with a direct bandgap with high tunability from ~0.3 eV (bulk) to ~2.0 eV (single-layer) and hole mobilities above 10,000 $cm^2V^{-1}s^{-1}$ [1–9]. Researchers have been striving to create enlarged band gaps in graphene, but with little success. On the other hand, the advent of the semiconducting BP was a breakthrough in the field [3, 4, 10]. In 2014, the first field-effect transistor (FET) embracing micrometer-sized plates of few-layer BP was presented [8]. Single-layer and few-layer BP materials also show some other fantastic features, such as highly anisotropic transport properties, excellent thermoelectric and optical responses, negative Poisson's ratio, strain-modulated conduction bands, and promising mechanical properties [11]. The unique properties of BP have triggered a plethora of research in this area, which holds appreciable prospects for the design of plenty of new devices in many growing fields [2, 6, 12]. This chapter is intended to provide a summary of the present status of the field, which follows by the opportunities, present challenges, and likely future directions of the BP materials.

2 Summary of the Present Status

The previous chapters presented the state-of-the-art research and progress in the field of BP materials such as phosphorene, phosphorene-based nanotubes and nanoribbons, BP-based fullerene-like nanostructures, few-layer and multilayer BP, and van der Waals (vdW) heterostructures containing phosphorene or few-layer BP. Their

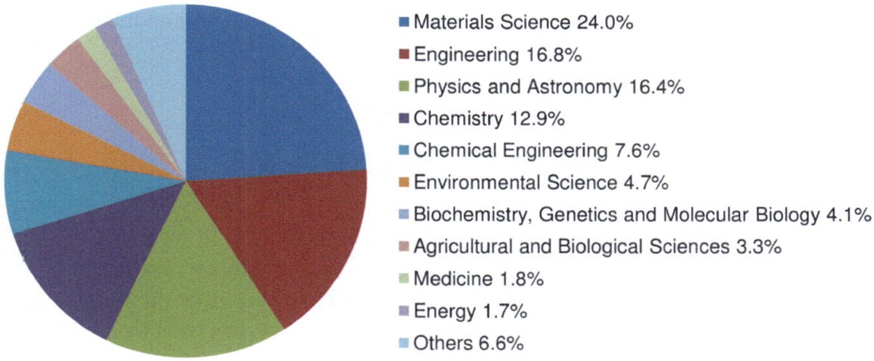

Fig. 1 Publications on BP leading to 2019 by subject area (according to Scopus)

geometries, properties, and applications were discussed in detail. Overall, the exceptional characteristics of BP, such as high charge-carrier mobility, direct bandgap, strong in-plane anisotropy, high conductivity, great lateral flexibility, and relatively high mechanical strength turn it into a promising 2D platform for a wide range of applications [5, 6, 12–15]. Interestingly, various characteristics of BP can be regulated using functionalization, chemical doping, atomic/molecular decoration, nanostructuring, induced electric field, elastic strain, and hybridization with different substrates [6, 14–16]. Benefiting from the broad tunability of its intriguing features, BP can be an ideal choice for different applications in various electronic devices, optics, sensors, drug delivery, bio-imaging, adsorption, batteries, solar cells, radio-frequency devices, and catalysis [5, 6, 10, 12, 14, 15, 17–19]. Researchers from various disciplines have been engaged in extensive research conducted to date on BP. Figure 1 depicts the pie chart of different fields of science and engineering in which the BP reports have appeared. The most involved majors include materials science, engineering, physics, and chemistry. However, the exotic properties and tenability of BP have spurred research in almost all subject areas.

The high structural and functional anisotropy of BP with respect to the electronic and thermal behaviors can be utilized for the fabrication of robust BP-containing thermoelectric devices [5, 14, 20, 21]. Moreover, the high mechanical strength of BP makes it an appropriate candidate as a reinforcing filler for application in composite materials [14, 22].

A black phosphorene nanoribbon (PNR) with zigzag-aligned crystal edges behaves as a metal with antiferromagnetic spin ordering. On the contrary, the PNR with armchair-directed edges remains a semiconductor [23]. Moreover, the band gaps of these nanostructures are tunable through the exposure to external field [24], strain [25], chemical functionalization, and defect passivation [26]. Although the PNRs are not as strong as a graphene-based nanoribbon (GNR) and the counterparts based on hexagonal boron nitride (h-BN) and transition metal dichalcogenide (TMD), the in-plane rigidity of a PNR is still substantially higher compared to conventional solids [14, 27]. Phosphorene-based fullerene-like structures have also received a great deal of interest. Unlike the MoS_2 fullerenes, these nanoparticles are semiconducting materials [14].

3 Opportunities and Future Challenges

The number of annual publications on BP has been increasing in recent years, with a rate exceeding other 2D materials [5]. Figure 2 represents this massive expanding growth in the publications before 2019. A sudden takeoff is evident in the research activities from 2014 on. The authorship of these publications has been from different countries, mainly China and the United States (Fig. 3). According to the trend shown in Fig. 2, the research in this area is still far from the plateau of productivity and, hence, the research activities may continue.

Despite the remarkable progress and the extensive studies in the exploration of phosphorene and BP-based nanomaterials, there are still limitations and the research

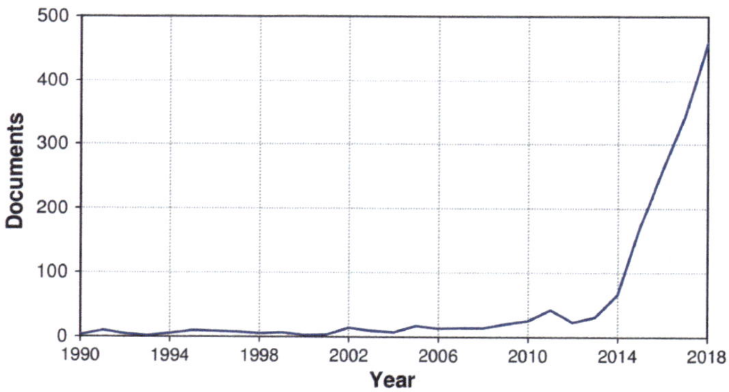

Fig. 2 Number of documents published per year on BP (according to Scopus)

Fig. 3 Published BP-related documents per country (according to Scopus)

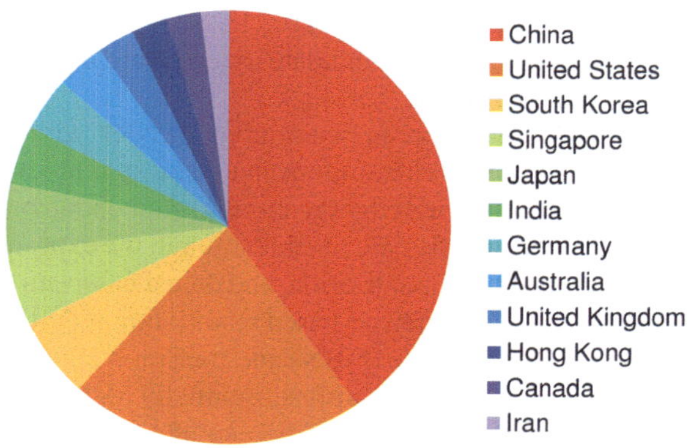

status in this area can be still regarded to be at its early stage relative to the more well-studied carbon and TMD nanostructures, such that several fundamental and applied aspects are still in need of further determination [6, 14, 15].

For instance, although the methodologies involving high pressure and catalyst have shown to be quite useful, there is still an urgent need in the developments of facile preparation methods for high-quality wafer-scale BP thin films with strict control over the thickness to nourish the future technology [5, 6, 14, 21]. A practical method for truly convenient and robust transfer of the BP nanosheets from one substrate to another is lacking [14]. The same lack of large-scale synthesis methods explains why the experimental studies of thermoelectric devices, band structure, magnetism, and optical devices have been rather scarce [5]. Inspired from the success of chemical vapor deposition (CVD) in the bottom-up synthesis of graphene, TMDs, and even other reactive monolayers such as silicene, germanene, and stanene, the CVD approach may be a possible avenue for scalable fabrication of uniform and high-quality phosphorene in the future [5, 6]. For this purpose, its fragile and active surface is to be circumvented, and efficient substrates should be found as the two prime concerns against phosphorene chemical growth [5]. In most of the previous studies, the few-layer BP has been isolated mechanically from the bulk-phase BP [6], which is not cost-effective [5]. Recent findings on the growth of high-purity crystallites of orthorhombic BP type from a red phosphorus (RP) phase using Sn/SnI_4 has sparked much hope for new possibilities [4, 5]. A great advantage of this pathway is that the RP precursor is more abundant and relatively inexpensive [5].

Recently, however, the liquid-phase exfoliation of BP monolayer in several solvents was proposed theoretically [15, 28]. The molecular dynamics (MD) simulations have demonstrated that if the chosen solvent is of a planar structure and both the solvent intermolecular cohesive forces as well as the adhesive forces between the BP monolayer and solvent molecules are considerably strong, the dispersed exfoliation of the BP nano-flakes will be favorable [15]. Exploring appropriate solvents may thus expectedly help accomplish the large-scale production of BP nanostructures through liquid-phase exfoliation in the near future [15].

Whereas bulk BP crystals are reasonably stable over the long term, the single-layer sample degrades within hours [29, 30]. The oxidative degradation occurs in the presence of O_2 and other oxide fragments of low molecular weight [3, 12, 31–33]. The same oxidation further makes the surface more sensitive to humidity. The susceptibility to the exothermic reaction with water turns the top layer of BP into hydrated phosphorus oxides, which is a negative factor in the electronic devices [3, 12, 18, 34]. To alleviate or impede the oxidative degradation, a non-covalent coating of the BP nanostructures with small electron-withdrawing molecules [35] would help [30]. The established procedures, such as low-temperature atomic layer deposition (ALD) of a proper dielectric cover such as alumina, PMMA encapsulation, liquid-phase exfoliation using anhydrous organic solvents such as N-methylpyrrolidone (NMP), and encapsulation in 2D stacked heterostructures have shown to be effective [3, 6, 36]. Some other suggested routes include the encapsulation with other suitable 2D nanomaterials, e.g. graphene or h-BN [33] and surface functionalization with titania, polyimide, aryl diazonium, or titanium sulfonate [12, 30, 36, 37].

Covalent functionalization can substantially help in this regard. Potential functional groups are hydroxyl group (–OH), carboxyl group (–COOH), amino group (–NH_2), and so forth [6, 38]. BP functionalization with 4-azidobenzoic acid [32], BH_3 [39], aryl groups [40], azodiisobutyronitrile (AIBN) [41], polydimethyldiallyl ammonium chloride (PDDA) [42], aromatic molecules (e.g., benzene and anthracene) [43], and more complicated systems [44] have recently been reported as well. Evidently, more eco-friendly and less-costly precursors are desired for large-scale implementation [6]. Nevertheless, the same sensitivity to water vapor and the reactivity to oxygen can be utilized to the advantage of gas sensing and to achieve tunable optical and excitonic properties [18, 45].

Substitutional doping of different heterospecies in many theoretical studies has shown to be quite useful for improving the electronic, optoelectronic, and magnetic behaviors of BP [6]. Perhaps, the most systematic research on BP modification has been implemented in this area [46–50]. Such improved systems have been predicted to perform satisfactorily for adsorption, detection, and catalysis applications [12]. Examples of target molecules include but are not limited to CO [48, 51], CO_2 [52] H_2 [53], NO [18], NO_2 [18], H_2S [54], NH_3 [54], H_2O_2 [18, 55], HCN [54], and heavy metals [18, 56]. Metal-doped phosphorene samples are also considered as potential candidates for N_2, H_2O, SO_2, phenylacetylene, and nitrophenol catalysis [14, 18, 57–63]. Despite the tremendous interest in this area, there is still an ample room and a promising outlook for future research, particularly with respect to sensing and catalysis.

Further experimental verifications are still required in which the strategies with control over concentrations and distributions are of high importance [6]. A precise knowledge of the atomic structures and the electronic behavior of the improved nanostructures is crucial for insights into the structure–function relationships. Meanwhile, characterization data from advanced methods, e.g. the scanning tunnelling microscopy (STM) and the aberration-corrected scanning transmission electron microscopy (STEM) can be insightful along the fundamental investigations using density-functional theory (DFT) computations and MD simulations. Due to their functional role and tremendous contribution to an in-depth understanding of the physicochemical phenomena, the theoretical interpretations and predictions will retain their place in parallel with the experimental studies of modified BP materials [6].

Sill of great interest is further exploration of hybrid BP nanomaterials using more 2D nanomaterials, such as graphene oxide, graphdiyne, metal oxides (e.g., WO_3, V_2O_5), TMDs (e.g., WS_2, WSe_2), metal–organic frameworks (MOFs), layered metal hydroxides (LMHs), and more [6]. Both configuration and sequence of stacking are plausible degrees of freedom for the BP modification [6].

A balanced combination of different modification strategies can also render as a brand-new avenue to fabricate multipurpose BP nanostructures with unprecedented performances [6]. To this end, the influence of several factors is at least quantitatively unknown. Notably, the effect of lattice imperfections (defects and vacancies) and their physicochemical (electronic, photonic, and mechanical) ramifications in the low-dimensional BP nanostructures should be further understood. Therefore, the emerging trend of novel BP doped/functional nanomaterials is expected to go on to thrive with broad implications at the interfaces of different disciplines, such as chemistry, physics, biology, materials science, and engineering [4, 6, 10]. Following the capabilities of the BP-based vdW heterostructures [64], the development and properties of superior devices prepared by assembling different 2D materials, e.g. the highly conductive zero-bandgap graphene [33], the insulating h-BN [65], and the semiconducting BP constitute an attractive area for the present research on 2D materials [6, 10, 12, 14, 19, 21, 66, 67].

One of the unique features that sets BP aside from other 2D nanomaterials is regarded to be its polymorphism. As discussed in the previous chapters, other 2D phases of P, e.g. the blue phosphorus allotrope, have been predicted, being virtually degenerate while chemically reactive [68]. Such predictions promote our understanding of BP chemistry. However, there is a great challenge of how to realize these allotropes and their combinations under controlled preparation conditions. To name a few, postulated phosphorene nanotubes and fullerene analogues are also yet to be fabricated. With this regard, the experimental protocols for effective control of their diameter, number of shells, and chirality would be highly desirable [14, 15].

Despite the research works which the recent years have witnessed, more efficient procedures are still to be presented for the passivation and functionalization of phosphorene monolayer, BP nanotubes, PNRs, and few-layer BP flakes. The possibility and mechanism of crystalline growth of phosphorene-based nanostructures are attractive fundamentally. The study of the interactions between the single-layer or few-layer BP nanosheets and the supporting substrates equally deserve a great deal of attention [14]. Similarly, benefiting from the dispersion forces between the hexagonal rings of the BP surface and the aromatic/aliphatic guest molecules, BP can serve as an outstanding platform to molecularly self-assemble and epitaxially grow various thin films with attractive nanopatterns for application in optoelectronic devices, such as organic light emitting diodes (OLEDs), organic FET devices (OFETs), and organic photovoltaic materials (OPVs) [12, 69].

As pointed out previously, the primary application area for the BP nanostructures has been in the electronic devices. Nonetheless, the robust and reliable integration of phosphorene with other 2D nanomaterials to launch practically improved devices should remain an active field of research [3, 6, 14]. Compared to conventional heterojunctions, often prepared via molecular beam epitaxy, the 2D heterojunctions benefit from the controllability of the vertical thickness of every layer at the atomic scale [3].

Fortunately, the direct band-gap semiconducting behavior, wide-range light absorption ability, and high charge carrier mobility of phosphorene have already been confirmed experimentally [15]. Owing to the broad absorption spectrum spanned from visible light to infrared (IR), the optoelectronic potential of BP would be even beyond the capability of the monolayer TMD materials [3, 4].

From a fundamental point of view, black phosphorene can render as an intriguing model structure to investigate new physics phenomena. Notably, it is very attractive to explore different types of direction-dependent polaritons, such as excitons and plasmons. It was discussed in the previous chapters that the effective masses of the electric charge carriers in the armchair direction are about one-tenth of the corresponding values in the zigzag one. The strong electronic anisotropy of the BP nanostructures have also implied that the electrons and holes could be envisaged as classical subatomic particles in the zigzag direction, but behaving as relativistic particles in the armchair one. This peculiarity serves as a motivation to investigate the associated principles [14].

Over the past decade, 2D nanomaterials have attracted enormous attention as efficient supports in catalytic reactions, including water splitting, O_2 reduction, and CO_2 activation. Analogously, it is expected that the pristine and doped BP nanostructures operate satisfactorily for such applications, particularly with a single atom catalyst (SAC) concept [12, 70]. BP displays a wide electrochemical cathodic window because it does not present high electrocatalytic nature for hydrogen evolution. On the contrary, the anodic window of BP is rather limited because of its electrochemical oxidation ability at neutral pH [18, 71]. Owing to these properties, BP has also been envisaged to be a favorable energy storage substance for both Na- and Li-ion batteries with high specific capacities [12, 14, 72]. As pointed out previously, the research in such areas is still in its infancy. Different doping elements are expected to tailor the properties of pristine phosphorene for different catalytic reactions.

Hitherto, almost all applications of black phosphorene were demonstrated in the academic institutes and research laboratories. Nearly all of the applications introduced previously depend on the ambient stability of the BP nanostructures for any robustness in applicability [12, 14]. Substantial roughening after the Scotch tape-based exfoliation of the BP thin films has been reported based on the atomic force microscopy (AFM) images [73], which implied the instability of the fabricated sample [15]. Another study utilized the laser pruning approach to synthesize few-layer BP flakes under ambient conditions [74]. However, the laser applied could accelerate the oxidation as well [15]. Theoretical studies have also shown that black phosphorene is easily etched by oxygen [15, 75]. Therefore, to commercially apply the BP-related materials to the electronic devices, sensors, etc., it will be inevitable to conquer the (structural and chemical) degradation issue in future studies [12, 14, 15].

As inspired from the protection of the few-layer MoS_2 composites with graphene against restacking, black phosphorene/MoS_2 heterojunction p–n diodes with vdW interactions have been investigated within the same concept [4, 10, 15, 19, 76]. Other investigations have also provided encouraging data for encapsulation and heterostructuring of phosphorene with graphene or h-BN through interlayer noncovalent bonding [15, 77]. Hence, the FETs with encapsulated thin-layer BP cores with hysteresis-free transport properties remain stable over two weeks [15]. More efficient methods for the encapsulation of phosphorene monolayers are yet required [14]. Therefore, the developments of well-defined 2D phosphorene-based heterostructures are expected to be continued in future to effectively suppress the ambient degradability of BP nanostructures [6, 12, 15].

As phosphorus is an essential element in the living cells, several biomedical applications of BP can be still investigated, including the biomolecular interactions, drug delivery, and cell imaging [12]. Although typically regarded to be nontoxic for biosensing [16, 6, 18], the degree of toxicity of such nanomaterials needs more stringent explorations [12]. Moreover, the BP-based biosensors can be applied to the DNA sensing on the basis of the π–π stacking and hence, the hydrophobic interactions [12].

Overall, BP and particularly the single-layer nanostructure phosphorene as a rising family of 2D nanomaterials presents both challenges and opportunities for all material scientists including chemists, physicists, and biologists from both fundamental and applied points of view [14]. Meanwhile, both experimental and computational insights cooperate synergistically to provide a better future for the developments in this area.

References

1. Fei, R., Yang, L.: Strain-engineering the anisotropic electrical conductance of few-layer black phosphorus. Nano Lett. **14**(5), 2884–2889 (2014). https://doi.org/10.1021/nl500935z
2. Zhang, S., Guo, S., Chen, Z., Wang, Y., Gao, H., Gómez-Herrero, J., Ares, P., Zamora, F., Zhu, Z., Zeng, H.: Recent progress in 2D group-VA semiconductors: from theory to experiment. Chem. Soc. Rev. **47**(3), 982–1021 (2018). https://doi.org/10.1039/c7cs00125h
3. Liu, H., Neal, A.T., Zhu, Z., Luo, Z., Xu, X., Tománek, D., Ye, P.D.: Phosphorene: an unexplored 2D semiconductor with a high hole mobility. ACS Nano **8**(4), 4033–4041 (2014). https://doi.org/10.1021/nn501226z
4. Liu, H., Du, Y., Deng, Y., Ye, P.D.: Semiconducting black phosphorus: synthesis, transport properties and electronic applications. Chem. Soc. Rev. **44**(9), 2732–2743 (2015). https://doi.org/10.1039/c4cs00257a
5. Akhtar, M., Anderson, G., Zhao, R., Alruqi, A., Mroczkowska, J.E., Sumanasekera, G., Jasinski, J.B.: Recent advances in synthesis, properties, and applications of phosphorene. npj 2D Mater. Appl. **1**(1), 5 (2017). https://doi.org/10.1038/s41699-017-0007-5
6. Lei, W., Liu, G., Zhang, J., Liu, M.: Black phosphorus nanostructures: recent advances in hybridization, doping and functionalization. Chem. Soc. Rev. **46**(12), 3492–3509 (2017). https://doi.org/10.1039/c7cs00021a
7. Qiao, J., Kong, X., Hu, Z.-X., Yang, F., Ji, W.: High-mobility transport anisotropy and linear dichroism in few-layer black phosphorus. Nat. Commun. **5**, 4475 (2014). https://doi.org/10.1038/ncomms5475
8. Li, L., Yu, Y., Ye, G.J., Ge, Q., Ou, X., Wu, H., Feng, D., Chen, X.H., Zhang, Y.: Black phosphorus field-effect transistors. Nat. Nanotechnol. **9**, 372–377 (2014). https://doi.org/10.1038/nnano.2014.35
9. Liu, Q., Zhang, X., Abdalla, L.B., Fazzio, A., Zunger, A.: Switching a normal insulator into a topological insulator via electric field with application to phosphorene. Nano Lett. **15**(2), 1222–1228 (2015). https://doi.org/10.1021/nl5043769
10. Ren, J., Zhang, C., Li, J., Guo, Z., Xiao, H., Zhong, J.: Strain engineering of magnetic state in vacancy-doped phosphorene. Phys. Lett. A **380**(40), 3270–3277 (2016). https://doi.org/10.1016/j.physleta.2016.07.055
11. Pumera, M., Sofer, Z.: 2D monoelemental arsenene, antimonene, and bismuthene: beyond black phosphorus. Adv. Mater. **29**(21), 1605299 (2017). https://doi.org/10.1002/adma.201605299
12. Chowdhury, C., Datta, A.: Exotic physics and chemistry of two-dimensional phosphorus: phosphorene. J. Phys. Chem. Lett. **8**(13), 2909–2916 (2017). https://doi.org/10.1021/acs.jpclett.7b01290
13. Ling, X., Wang, H., Huang, S., Xia, F., Dresselhaus, M.S.: The renaissance of black phosphorus. Proc. Natl. Acad. Sci. **112**(15), 4523 (2015). https://doi.org/10.1073/pnas.1416581112

14. Sorkin, V., Cai, Y., Ong, Z., Zhang, G., Zhang, Y.W.: Recent advances in the study of phosphorene and its nanostructures. Crit. Rev. Solid State Mater. Sci. **42**(1), 1–82 (2017). https://doi.org/10.1080/10408436.2016.1182469
15. Jing, Y., Zhang, X., Zhou, Z.: Phosphorene: what can we know from computations? WIREs Comput. Mol. Sci. **6**(1), 5–19 (2016). https://doi.org/10.1002/wcms.1234
16. Kou, L., Chen, C., Smith, S.C.: Phosphorene: fabrication, properties, and applications. J. Phys. Chem. Lett. **6**(14), 2794–2805 (2015). https://doi.org/10.1021/acs.jpclett.5b01094
17. Kulish, V.V., Malyi, O.I., Persson, C., Wu, P.: Phosphorene as an anode material for Na-ion batteries: a first-principles study. Phys. Chem. Chem. Phys. **17**(21), 13921–13928 (2015). https://doi.org/10.1039/c5cp01502b
18. Pumera, M.: Phosphorene and black phosphorus for sensing and biosensing. TrAC Trends Anal. Chem. **93**, 1–6 (2017). https://doi.org/10.1016/j.trac.2017.05.002
19. Çakır, D., Sahin, H., Peeters, F.M.: Tuning of the electronic and optical properties of single-layer black phosphorus by strain. Phys. Rev. B **90**(20), 205421 (2014)
20. Zhang, G., Zhang, Y.-W.: Strain effects on thermoelectric properties of two-dimensional materials. Mech. Mater. **91**, 382–398 (2015). https://doi.org/10.1016/j.mechmat.2015.03.009
21. Du, Y., Luo, Z., Liu, H., Xu, X., Ye, P.D.: Anisotropic properties of black phosphorus. In: Avouris, P., Low, T., Heinz, T.F. (eds.) 2D Materials: Properties and Devices, pp. 413–434. Cambridge University Press, Cambridge (2017). https://doi.org/10.1017/9781316681619.023
22. Hanlon, D., Backes, C., Doherty, E., Cucinotta, C.S., Berner, N.C., Boland, C., Lee, K., Harvey, A., Lynch, P., Gholamvand, Z., Zhang, S., Wang, K., Moynihan, G., Pokle, A., Ramasse, Q.M., McEvoy, N., Blau, W.J., Wang, J., Abellan, G., Hauke, F., Hirsch, A., Sanvito, S., O'Regan, D.D., Duesberg, G.S., Nicolosi, V., Coleman, J.N.: Liquid exfoliation of solvent-stabilized few-layer black phosphorus for applications beyond electronics. Nat. Commun. **6**, 8563 (2015). https://doi.org/10.1038/ncomms9563
23. Zhang, J., Liu, H.J., Cheng, L., Wei, J., Liang, J.H., Fan, D.D., Shi, J., Tang, X.F., Zhang, Q.J.: Phosphorene nanoribbon as a promising candidate for thermoelectric applications. Sci. Rep. **4**, 6452 (2014). https://doi.org/10.1038/srep06452
24. Taghizadeh Sisakht, E., Zare, M.H., Fazileh, F.: Scaling laws of band gaps of phosphorene nanoribbons: a tight-binding calculation. Phys. Rev. B **91**(8), 085409 (2015). https://doi.org/10.1103/PhysRevB.91.085409
25. Han, X., Stewart, H.M., Shevlin, S.A., Catlow, C.R.A., Guo, Z.X.: Strain and orientation modulated bandgaps and effective masses of phosphorene nanoribbons. Nano Lett. **14**(8), 4607–4614 (2014). https://doi.org/10.1021/nl501658d
26. Guo, H., Lu, N., Dai, J., Wu, X., Zeng, X.C.: Phosphorene nanoribbons, phosphorus nanotubes, and van der Waals multilayers. J. Phys. Chem. C **118**(25), 14051–14059 (2014). https://doi.org/10.1021/jp505257g
27. Sorkin, V., Zhang, Y.W.: The structure and elastic properties of phosphorene edges. Nanotechnology **26**(23), 235707 (2015). https://doi.org/10.1088/0957-4484/26/23/235707
28. Sresht, V., Pádua, A.A.H., Blankschtein, D.: Liquid-phase exfoliation of phosphorene: design rules from molecular dynamics simulations. ACS Nano **9**(8), 8255–8268 (2015). https://doi.org/10.1021/acsnano.5b02683
29. Gusmão, R., Sofer, Z., Pumera, M.: Black phosphorus rediscovered: from bulk material to monolayers. Angew. Chem. Int. Ed. **56**(28), 8052–8072 (2017). https://doi.org/10.1002/anie.201610512
30. Hirsch, A., Hauke, F.: Post-graphene 2D chemistry: the emerging field of molybdenum disulfide and black phosphorus functionalization. Angew. Chem. Int. Ed. **57**(16), 4338–4354 (2018). https://doi.org/10.1002/anie.201708211
31. Pei, J., Gai, X., Yang, J., Wang, X., Yu, Z., Choi, D.-Y., Luther-Davies, B., Lu, Y.: Producing air-stable monolayers of phosphorene and their defect engineering. Nat. Commun. **7**, 10450 (2016). https://doi.org/10.1038/ncomms10450
32. Liu, Y., Gao, P., Zhang, T., Zhu, X., Zhang, M., Chen, M., Du, P., Wang, G.-W., Ji, H., Yang, J., Yang, S.: Azide passivation of black phosphorus nanosheets: covalent functionalization affords ambient stability enhancement. Angew. Chem. Int. Ed. **58**(5), 1479–1483 (2019). https://doi.org/10.1002/anie.201813218

33. Doganov, R.A., O'Farrell, E.C.T., Koenig, S.P., Yeo, Y., Ziletti, A., Carvalho, A., Campbell, D.K., Coker, D.F., Watanabe, K., Taniguchi, T., Neto, A.H.C., Özyilmaz, B.: Transport properties of pristine few-layer black phosphorus by van der Waals passivation in an inert atmosphere. Nat. Commun. **6**, 6647 (2015). https://doi.org/10.1038/ncomms7647
34. Wang, G., Slough, W.J., Pandey, R., Karna, S.P.: Degradation of phosphorene in air: understanding at atomic level. 2D Mater. **3**(2), 025011 (2016). https://doi.org/10.1088/2053-1583/3/2/025011
35. Abellán, G., Lloret, V., Mundloch, U., Marcia, M., Neiss, C., Görling, A., Varela, M., Hauke, F., Hirsch, A.: Noncovalent functionalization of black phosphorus. Angew. Chem. Int. Ed. **55**(47), 14557–14562 (2016). https://doi.org/10.1002/anie.201604784
36. Carvalho, A., Wang, M., Zhu, X., Rodin, A.S., Su, H., Castro Neto, A.H.: Phosphorene: from theory to applications. Nat. Rev. Mater. **1**, 16061 (2016). https://doi.org/10.1038/natrevmats.2016.61
37. Ryder, C.R., Wood, J.D., Wells, S.A., Yang, Y., Jariwala, D., Marks, T.J., Schatz, G.C., Hersam, M.C.: Covalent functionalization and passivation of exfoliated black phosphorus via aryl diazonium chemistry. Nat. Chem. **8**, 597 (2016). https://doi.org/10.1038/nchem.2505
38. Shao, L., Sun, H., Miao, L., Chen, X., Han, M., Sun, J., Liu, S., Li, L., Cheng, F., Chen, J.: Facile preparation of NH_2-functionalized black phosphorene for the electrocatalytic hydrogen evolution reaction. J. Mater. Chem. A **6**(6), 2494–2499 (2018). https://doi.org/10.1039/c7ta10884b
39. Ienco, A., Manca, G., Peruzzini, M., Mealli, C.: Modelling strategies for the covalent functionalization of 2D phosphorene. Dalton Trans. **47**(48), 17243–17256 (2018). https://doi.org/10.1039/c8dt03628d
40. van Druenen, M., Davitt, F., Collins, T., Glynn, C., O'Dwyer, C., Holmes, J.D., Collins, G.: Covalent functionalization of few-layer black phosphorus using iodonium salts and comparison to diazonium modified black phosphorus. Chem. Mater. **30**(14), 4667–4674 (2018). https://doi.org/10.1021/acs.chemmater.8b01306
41. Hu, H., Gao, H., Gao, L., Li, F., Xu, N., Long, X., Hu, Y., Jin, J., Ma, J.: Covalent functionalization of black phosphorus nanoflakes by carbon free radicals for durable air and water stability. Nanoscale **10**(13), 5834–5839 (2018). https://doi.org/10.1039/c7nr06085h
42. Feng, Q., Liu, H., Zhu, M., Shang, J., Liu, D., Cui, X., Shen, D., Kou, L., Mao, D., Zheng, J., Li, C., Zhang, J., Xu, H., Zhao, J.: Electrostatic functionalization and passivation of water-exfoliated few-layer black phosphorus by poly dimethyldiallyl ammonium chloride and its ultrafast laser application. ACS Appl. Mater. Interfaces **10**(11), 9679–9687 (2018). https://doi.org/10.1021/acsami.8b00556
43. Scotognella, F., Kriegel, I., Sassolini, S.: Covalent functionalized black phosphorus quantum dots. Opt. Mater. **75**, 521–524 (2018). https://doi.org/10.1016/j.optmat.2017.11.016
44. Cao, Y., Tian, X., Gu, J., Liu, B., Zhang, B., Song, S., Fan, F., Chen, Y.: Covalent functionalization of black phosphorus with conjugated polymer for information storage. Angew. Chem. Int. Ed. **57**(17), 4543–4548 (2018). https://doi.org/10.1002/anie.201712675
45. Sadki, S., Drissi, L.B.: Tunable optical and excitonic properties of phosphorene via oxidation. J. Phys.: Condens. Matter **30**(25), 255703 (2018). https://doi.org/10.1088/1361-648x/aac403
46. Sun, H., Shang, Y., Yang, Y., Guo, M.: Realization of N-type semiconducting of phosphorene through surface metal doping and work function study. J. Nanomater. **2018**, 9 (2018). https://doi.org/10.1155/2018/6863890
47. Wang, K., Wang, H., Zhang, M., Zhao, W., Liu, Y., Qin, H.: The electronic and magnetic properties of multi-atom doped black phosphorene. Nanomaterials **9**(2), 311 (2019). https://doi.org/10.3390/nano9020311
48. Sun, X., Luan, S., Shen, H., Lei, S.: Effect of metal doping on carbon monoxide adsorption on phosphorene: a first-principles study. Superlattices Microstruct. **124**, 168–175 (2018). https://doi.org/10.1016/j.spmi.2018.09.037
49. Lei, S.Y., Luan, S., Yu, H.: Co-doped phosphorene: enhanced sensitivity of CO gas sensing. Int. J. Mod. Phys. B **32**(06), 1850068 (2018). https://doi.org/10.1142/s0217979218500686

50. Musle, V., Choudhary, S.: Tuning the optical properties of phosphorene by adsorption of alkali metals and halogens. Opt. Quant. Electron. **50**(7), 285 (2018). https://doi.org/10.1007/s11082-018-1548-3
51. Lei, S.Y., Yu, Z.Y., Shen, H.Y., Sun, X.L., Wan, N., Yu, H.: CO adsorption on metal-decorated phosphorene. ACS Omega **3**(4), 3957–3965 (2018). https://doi.org/10.1021/acsomega.8b00133
52. Zhang, H.-p., Du, A., Shi, Q.-b., Zhou, Y., Zhang, Y., Tang, Y.: Adsorption behavior of CO_2 on pristine and doped phosphorenes: a dispersion corrected DFT study. J. CO_2 Utilization **24**, 463–470 (2018). https://doi.org/10.1016/j.jcou.2018.02.005
53. Zhang, H.-p., Hu, W., Du, A., Lu, X., Zhang, Y.-p., Zhou, J., Lin, X., Tang, Y.: Doped phosphorene for hydrogen capture: a DFT study. Appl. Surf. Sci. **433**(Supplement C), 249–255 (2018). https://doi.org/10.1016/j.apsusc.2017.09.243
54. Kuang, A., Ran, Y., Peng, B., Kuang, M., Wang, G., Yuan, H., Tian, C., Chen, H.: Adsorption and decomposition of metal decorated phosphorene toward H_2S, HCN and NH_3 molecules. Appl. Surf. Sci. **473**, 242–250 (2019). https://doi.org/10.1016/j.apsusc.2018.12.131
55. Yan, S., Wang, B., Wang, Z., Hu, D., Xu, X., Wang, J., Shi, Y.: Supercritical carbon dioxide-assisted rapid synthesis of few-layer black phosphorus for hydrogen peroxide sensing. Biosens. Bioelectron. **80**(Supplement C), 34–38 (2016). https://doi.org/10.1016/j.bios.2016.01.043
56. Li, P., Zhang, D., Liu, J., Chang, H., Ye, Sun, Yin, N.: Air-stable black phosphorus devices for ion sensing. ACS Appl. Mater. Interfaces **7**(44), 24396–24402 (2015). https://doi.org/10.1021/acsami.5b07712
57. Wei, Z., Zhang, Y., Wang, S., Wang, C., Ma, J.: Fe-doped phosphorene for the nitrogen reduction reaction. J. Mater. Chem. A **6**(28), 13790–13796 (2018). https://doi.org/10.1039/c8ta03989e
58. Yang, Q., Meng, R.S., Jiang, J.K., Liang, Q.H., Tan, C.J., Cai, M., Sun, X., Yang, D.G., Ren, T.L., Chen, X.P.: First-principles study of sulfur dioxide sensor based on phosphorenes. IEEE Electron Dev. L **37**(5), 660–662 (2016). https://doi.org/10.1109/led.2016.2543243
59. Kakaei, K., Esrafili, M.D., Ehsani, A.: Introduction to catalysis. In: Kakaei, K., Esrafili, M.D., Ehsani, A. (eds.) Interface Science and Technology, vol. 27, pp. 1–21. Elsevier (2019). https://doi.org/10.1016/B978-0-12-814523-4.00001-0
60. Makhlouf, A.S.H., Tiginyanu, I.: Nanocoatings and Ultra-Thin Films: Technologies and Applications. Elsevier Science (2011)
61. Luo, Y., Ren, C., Wang, S., Li, S., Zhang, P., Yu, J., Sun, M., Sun, Z., Tang, W.: Adsorption of transition metals on black phosphorene: a first-principles study. Nanoscale Res. Lett. **13**(1), 282 (2018). https://doi.org/10.1186/s11671-018-2696-x
62. Caporali, M., Serrano-Ruiz, M., Telesio, F., Heun, S., Nicotra, G., Spinella, C., Peruzzini, M.: Decoration of exfoliated black phosphorus with nickel nanoparticles and its application in catalysis. Chem. Commun. **53**(79), 10946–10949 (2017). https://doi.org/10.1039/c7cc05906j
63. Zhang, L., Gao, L.-F., Li, L., Hu, C.-X., Yang, Q.-Q., Zhu, Z.-Y., Peng, R., Wang, Q., Peng, Y., Jin, J., Zhang, H.-L.: Negatively charged 2D black phosphorus for highly efficient covalent functionalization. Mater. Chem. Front. **2**(9), 1700–1706 (2018). https://doi.org/10.1039/c8qm00237a
64. Gong, K., Zhang, L., Ji, W., Guo, H.: Electrical contacts to monolayer black phosphorus: a first-principles investigation. Phys. Rev. B **90**(12), 125441 (2014). https://doi.org/10.1103/PhysRevB.90.125441
65. Chen, X., Wu, Y., Wu, Z., Han, Y., Xu, S., Wang, L., Ye, W., Han, T., He, Y., Cai, Y., Wang, N.: High-quality sandwiched black phosphorus heterostructure and its quantum oscillations. Nat. Commun. **6**, 7315 (2015). https://doi.org/10.1038/ncomms8315
66. Xia, F., Wang, H., Jia, Y.: Rediscovering black phosphorus as an anisotropic layered material for optoelectronics and electronics. Nat. Commun. **5**, 4458 (2014). https://doi.org/10.1038/ncomms5458
67. Dai, J., Zeng, X.C.: Bilayer phosphorene: effect of stacking order on bandgap and its potential applications in thin-film solar cells. J. Phys. Chem. Lett. **5**(7), 1289–1293 (2014). https://doi.org/10.1021/jz500409m

68. Wu, M., Fu, H., Zhou, L., Yao, K., Zeng, X.C.: Nine new phosphorene polymorphs with non-honeycomb structures: a much extended family. Nano Lett. **15**(5), 3557–3562 (2015). https://doi.org/10.1021/acs.nanolett.5b01041
69. Mukhopadhyay, T.K., Datta, A.: Ordering and dynamics for the formation of two-dimensional molecular crystals on black phosphorene. J. Phys. Chem. C **121**(18), 10210–10223 (2017). https://doi.org/10.1021/acs.jpcc.7b02480
70. Zhao, J., Liu, X., Chen, Z.: Frustrated Lewis Pair catalysts in two dimensions: B/Al-doped phosphorenes as promising catalysts for hydrogenation of small unsaturated molecules. ACS Catal. **7**(1), 766–771 (2017). https://doi.org/10.1021/acscatal.6b02727
71. Wang, L., Sofer, Z., Pumera, M.: Voltammetry of layered black phosphorus: electrochemistry of multilayer phosphorene. ChemElectroChem **2**(3), 324–327 (2015). https://doi.org/10.1002/celc.201402363
72. Li, W., Yang, Y., Zhang, G., Zhang, Y.-W.: Ultrafast and directional diffusion of lithium in phosphorene for high-performance lithium-ion battery. Nano Lett. **15**(3), 1691–1697 (2015). https://doi.org/10.1021/nl504336h
73. Koenig, S.P., Doganov, R.A., Schmidt, H., Neto, A.H.C., Özyilmaz, B.: Electric field effect in ultrathin black phosphorus. Appl. Phys. Lett. **104**(10), 103106 (2014). https://doi.org/10.1063/1.4868132
74. Lu, J., Wu, J., Carvalho, A., Ziletti, A., Liu, H., Tan, J., Chen, Y., Castro Neto, A.H., Özyilmaz, B., Sow, C.H.: bandgap engineering of phosphorene by laser oxidation toward functional 2D materials. ACS Nano **9**(10), 10411–10421 (2015). https://doi.org/10.1021/acsnano.5b04623
75. Dai, J., Zeng, X.C.: Structure and stability of two dimensional phosphorene with =O or =NH functionalization. RSC Adv. **4**(89), 48017–48021 (2014). https://doi.org/10.1039/c4ra02850c
76. Carvalho, A., Neto, A.H.C.: Phosphorene: overcoming the oxidation barrier. ACS Central Sci. **1**(6), 289–291 (2015). https://doi.org/10.1021/acscentsci.5b00304
77. Avsar, A., Vera-Marun, I.J., Tan, J.Y., Watanabe, K., Taniguchi, T., Castro Neto, A.H., Özyilmaz, B.: Air-stable transport in graphene-contacted, fully encapsulated ultrathin black phosphorus-based field-effect transistors. ACS Nano **9**(4), 4138–4145 (2015). https://doi.org/10.1021/acsnano.5b00289

Black Phosphorous Photodetectors

Hui Qiao, Chenguang Duan, Zongyu Huang and Xiang Qi

Abstract As a new two-dimensional material, black phosphorus (BP) has a direct band gap that can be controlled with the change of the number of layers, good light absorption rate and excellent electron mobility making it a good application in the field of photodetectors. To date, many studies on BP photodetectors have been reported and have demonstrated excellent photoresponsive performance. Here, we have a comprehensive summary and review of research reports on BP-based photodetectors. We first briefly summarize the structural characteristics, optical and electronic properties of BP. Then BP-based photodetectors are divided into three categories according to the structure of the material and the configuration of the device (BP photodetectors, BP heterojunction photodetectors, photoelectrochemical (PEC)-type BP photodetectors) for summary and description. Finally, the research status of BP-based photodetectors is summarized, and its future development direction is prospected. The content of this chapter can give readers a comprehensive understanding of BP-based photodetectors, and has certain guiding significance for the further development of photodetectors.

Keywords Black phosphorus · Photodetectors · Heterojunction · Photoelectrochemical

1 Introduction

The photodetector is a device that converts an optical signal into an electrical signal [1]. The working mechanism of photodetectors is based on photoelectric effect. As early as 1887, the German physicist Hertz discovered that under the irradiation of electromagnetic waves of a certain wavelength, the electrons inside the material are excited by photons to form a current, also known as photo-generated current. In the past 10 years, photodetectors have been continuously developed and widely used

H. Qiao · C. Duan · Z. Huang · X. Qi (✉)
Hunan Key Laboratory of Micro-Nano Energy Materials and Devices, Laboratory for Quantum Engineering and Micro-Nano Energy Technology, School of Physics and Optoelectronic, Xiangtan University, Hunan 411105, People's Republic of China
e-mail: xqi@xtu.edu.cn

Inamuddin et al. (eds.), *Black Phosphorus*, Engineering Materials,
https://doi.org/10.1007/978-3-030-29555-4_9

in many fields, such as optical communication [2], environmental monitoring [3], infrared sensors [4], laser range finder [5], etc. For photodetectors, the evaluation of their performance is very important. The main parameters include optical responsivity, response rate, and optical switching ratio [6]. In addition, the performance of photodetectors is affected and regulated by many factors, such as materials, irradiance, device configuration, electrode type, etc. Therefore, it is critical to find suitable photosensitive materials for the development of high performance photodetectors

Two-dimensional (2D) materials are a good candidate material widely cited in the field of photodetectors due to their excellent optical and electronic properties [7]. Recently, a new type of 2D monoelemental materials, black phosphorus (BP) nanosheets, has been successfully prepared and widely concerned. The bulk black phosphorus, which is the most stable allotrope of phosphorus, was prepared in 1914 and has a layered structure [8]. It has an adjustable direct band gap and the band gap is increased from 0.3 to 2.0 eV when the thickness of the bulk BP is reduced to a single-layer [9]. This narrow bandgap and broad-spectrum photoresponse behavior coverage the spectral range from the visible to near-infrared indicates that BP is a very promising candidate in the field of photodetectors [10]. Furthermore, Zhang and his team [11] used a scotch tape-based mechanical exfoliation method to separate few-layers of BP nanosheets from bulk BP and transfer them to a degraded doped silicon wafer covered with a thermally grown silicon dioxide layer to prepare field effect transistors. The charge carrier mobility of the transistor also exhibits a thickness dependence. When the thickness of the two-dimensional black phosphorus is 10 nm, the highest mobility value of 1000 $cm^2V^{-1}s^{-1}$ is obtained. In addition to high carrier mobility, the few-layers BP nanosheets has direct bandgap and strong light absorption efficiency, all of which indicate that BP has high performance photoresponsive performance. However the poor environmental stability severely limits its further development in practical applications. In order to solve this problem, many reports have proposed many ways to improve the stability of BP nanosheets [12]. Zhao et al. [13] designed a titanium sulfonate ligand (designated as TiL_4) to bind to a lone pair of electrons in BP, which can effectively inhibit the oxidation of BP and enhance its stability in air and water. Grasser et al. [14] used Al_2O_3 layer to package a black phosphorus field-effect transistor and found that the transistor exhibited stable device characteristics for at least 8 months. Besides this, the dispersion of BP nanosheets in some special solvents can also enhance their environmental stability, such as poly lactic-co-glycolic acid [15], N-methyl-2-pyrrolidone (NMP) [16], alkaline solution [17]. These research reports have greatly expanded the practical application of BP in the field of optoelectronics.

With the rapid development in recent years, the research on BP-based photodetectors is quite mature. To date, many photodetectors based on BP and BP heterojunctions have been reported. Therefore, it is very meaningful to conduct a comprehensive review report on the research report of BP-based photodetectors. However, so far, there is no systematic review of the BP-based photodetectors. In this section, we provide a comprehensive summary and description of photodetectors based on BP and BP heterojunctions. In the above, we first briefly summarize and review the optical properties, electronic properties and environmental stability of BP. Then, the latest research progress of BP-based photodetectors (Black Phosphorous-based Photodetectors, Black Phosphorous Heterojunction-Basde Photodetectors, and novel BP-based photoelectrochemical (PEC)-type photodetectors) is introduced in detail. Finally, the development status and research significance of BP-based photodetectors are summarized and discussed, and the further development direction of BP-based photodetectors is prospected. The content of this chapter has important guiding significance for the further development and application of black phosphorus, and also has guiding significance for the development and application of other new materials in the field of photodetectors.

2 Black Phosphorous-Based Photodetectors

2.1 Traditional Black Phosphorus-Based Photodetectors

As an important optoelectronic device, photodetectors have been widely used in military, industrial production and daily life. However, most semiconductor photodetectors can only detect ultraviolet and visible light due to their large band gap [7b, 18]. Fortunately, as a new type of monoelemental 2D semiconductor material, the emergence of BP has solved this problem very well. The adjustable band gap of BP with the change of the number of layers enables it to achieve broadband photoresponse behavior from visible light to infrared light range. In addition, BP also exhibits a direct band gap, high light absorption efficiency [19] and high carrier mobility [20], making BP a good candidate for photodetectors. So far, many studies on BP photodetectors have been reported [21]. Engel et al. [22] studied a photodetector based on multilayer of BP (Fig. 1a). The experimental results show that it can obtain high contrast images both in the visible ($\lambda = 532$ nm) and infrared ($\lambda = 1550$ nm) spectral ranges. In addition, the photodetector has an photoresponsivity of 20 and 5 mA/W at $\lambda = 532$ nm and $\lambda = 1550$ nm, respectively. Subsequently, Guo et al. [23] constructed a wide wavelength range photodetector ranging from 532 to 3.39 μm based on BP film with a thickness of about 10 nm (Fig. 1b). The device has low dark current and high photoconductivity gain. At 3.39 μm, the BP thin film photodetector has an photoresponsivity of up to 82 A/W. In addition have a good responsiveness to visible light and infrared light, BP photodetectors also have good responsiveness to ultraviolet light. Wu et al. [24] studied the photoresponse performance of field effect transistors based on a few-layers black phosphorus nanosheets (Fig. 1c) and found that the device can have broadband photoresponse characteristics from near ultraviolet to near infrared (from 310 to 950 nm). It is worth noting that the photoresponsiveness of FET in the near-ultraviolet region is 5 orders of magnitude higher than that in the visible-near-infrared region, and its UV photoresponsivity is ~9×10^4 AW^{-1} and the detectivity is ~3×10^{13} Jones. besides this, the photoresponse performance of the BP photodetector can be adjusted by its thickness due to its adjustable band gap as the number of layers changes. Nathan et al. [25] created

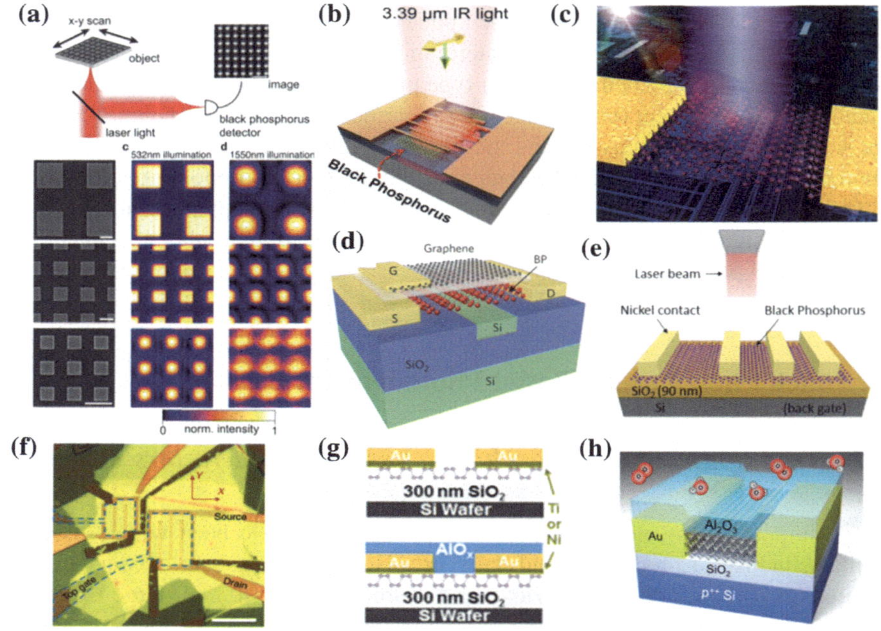

Fig. 1 **a** Schematic diagram of the imaging process of black phosphorus photodetector. **b** Black phosphorus mid-infrared photodetector configuration and illustration working at 3.39 μm. **c** Three-dimensional structure diagram of a few black phosphorus field effect transistors. **d** Structure diagram of a waveguide-integrated black phosphorus photodetector with few-layer graphene top-gate. **e** Schematic of the Broadband Black-Phosphorus Photodetectors device. **f** Structure of tunable BP mid-infrared photodetector based on dual-gate transistor configuration. **g** Schematic of a BP field effect transistor in an unpackaged and AlO_x package. **h** Schematic diagram of several layers of BP phototransistors with Al_2O_3 passivation layer. Figure 1a is reproduced with reference to 22 permission. Copyright 2014, American Chemical Society. Figure 1b is reproduced with reference to 23 permission. Copyright 2016, American Chemical Society. Figure 1c is reproduced with reference to 24 permission. Copyright 2015, American Chemical Society. Figure 1d is reproduced with reference to 25 permission. Copyright 2015, Macmillan Publishers Limited. Figure 1e is reproduced with reference to 26 permission. Copyright 2016, WILEY-VCH. Figure 1f is reproduced with reference to 21b permission. Copyright 2017, Springer Nature Publishing AG Fig. 1g is reproduced with reference to 28 permission. Copyright 2014, American Chemical Society Fig. 1h is reproduced with reference to 29 permission. Copyright 2017, IOP Publishing Ltd

a gated photodetector based on multilayer black phosphorus, which can operates with a very low dark currents at bias voltages (Fig. 1d). Under an optical power of 1.91 mW, the photodetectors with thicknesses of 11.5 and 100 nm achieve photoresponsivity of up to 135 and 657 mA W^{-1}, respectively. Furthermore, Huang et al. [26] used mechanical exfoliation to obtain a Few-layer BP flakes (the thickness of the BP flakes was 8 nm, corresponding to 15 layers) on a SiO_2 substrate (Fig. 1e), and constructed a high performance broadband BP photodetector with a wavelength range of 400–900 nm. The photoresponsivity of the photodetector reaches about 7

$\times$ 10^6 A W^{-1} at 20 K and 4.3 $\times$ 10^6 A W^{-1} at 300 K. This is the first study of the relationship between the photoresponsivity of few-layers BP photodetector and the temperature dependence. In addition, there are reports that the vertical electric field can dynamically adjust the optical response spectral range of the BP photodetector. Chen et al. [21b] successfully extended the photoresponse spectral range of a 3.7 μm thick BP photodetector to 7.7 μm using a vertical electric field (Fig. 1f). It is further proved that this widely tunable BP photodetector exhibits photoresponsivity of 518, 30 and 2.2 mA W^{-1} at 3.4, 5 and 7.7 μm, respectively. These reports indicate that several layers of BP are suitable for building high sensitivity broadband photoresponsive photodetectors due to their high mobility and small band gap.

However, 2D BP exposed to the environment is easily oxidized quickly, which greatly limits the further development and application of BP photodetectors [13, 27]. Therefore, a series of means have been proposed to improve the stability of BP photodetectors. Wood et al. [28] effectively inhibited the degradation of BP in the environment by depositing an AlO_x overlayers (Fig. 1g). In the experiment, the BP FET packaged with AlO_x film can maintain a high on/off ratio of ~10^3 and a mobility of ~100 $cm^2V^{-1}s^{-1}$ over 2 weeks under ambient conditions. In addition, Na et al. [29] successfully fabricated a small layer of BP-based phototransistors with good air stability through proper Al_2O_3 passivation (Fig. 1h). The device has a fast response time of less than 100 μs, low dark current of approximately 4 nA and a good optical response of ~6 mA W^{-1} at operating bias. It is worth noting that the device can operate stably in air for 6 months and still maintain the original optical response performance. Besides this, it can also enhance the stability of BP photodetectors by doping.

2.2 *Matel-Black Phosphorous Schottky Photodetectors*

In general, the design of a conventional photodetector requires the plating of two ohmic contact metal electrodes on the surface of the semiconductor material. When one of the ohmic contact electrolysis is replaced by a schottky barrier contact, an asymmetric schottky junction photodetector can be constructed. Metal-semiconductor schottky photodetectors have enhanced photoresponse performance compared to conventional ohmic contact photodetectors. This is due to the formation of a schottky barrier at the interface between the metal electrode and the semiconductor, which can promote efficient separation of photogenerated electron-hole pairs, thereby enhancing photoresponse performance. Therefore, constructing a metal-BP schottky junction is considered to be an effective means to enhance BP photoresponse performance. Miao et al. [30] constructed an asymmetric Au-BP-Al Schottky photodetector using Au and Al electrodes and evaluated its photoresponse performance (Fig. 2a). The BP schottky diode has a fast response of less than 2 ms, a responsiveness of $\approx$3.5 mA W^{-1}, and an external quantum efficiency of 0.65% at an incident light intensity of 38 Wcm^{-2}. Fast photoresponse rate is due to the built-in

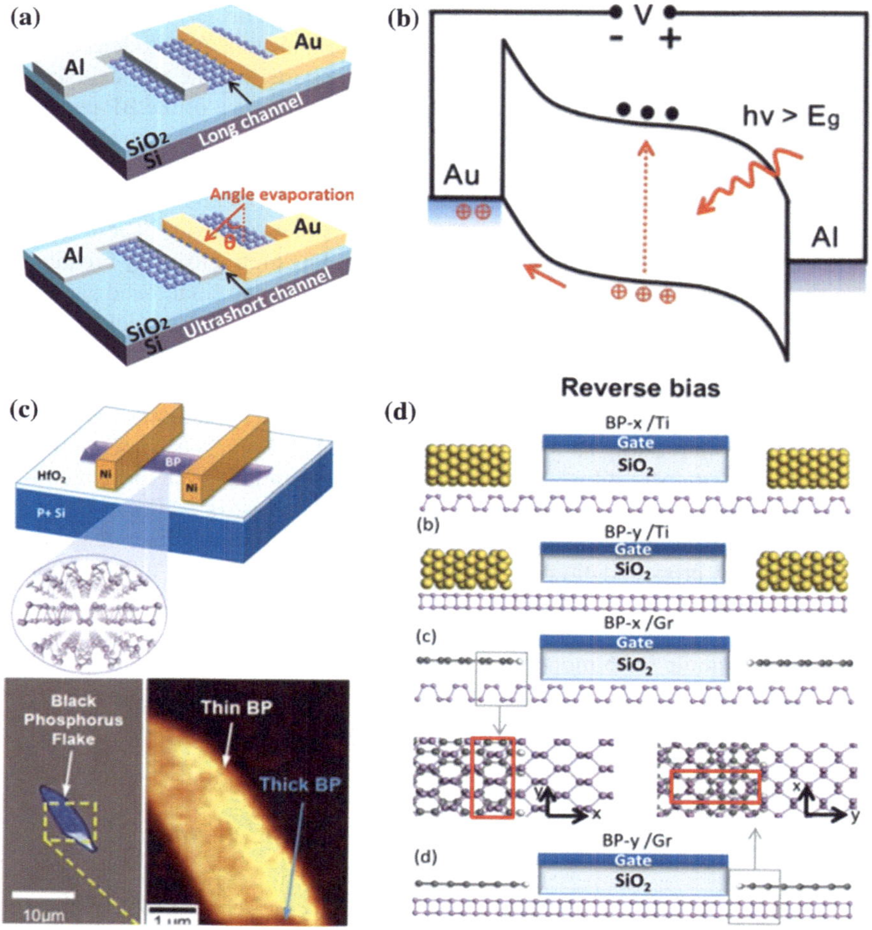

Fig. 2 **a** Schematic diagram of asymmetric Au-BP-Al Schottky photodetector with long channel (top) and short channels (down). **b** Energy band diagram of asymmetric Au-BP-Al Schottky diode and schematic diagram of photogenerated electron transfer. **c** Schematic diagram of low schottky barrier BP Field-Effect Devices. Illustration: AFM image of the edge of a sheet of BP. **d** Schematic diagram of BP schottky barrier transistor using graphene and Ti as electrodes Fig. 2a are reproduced with reference to 30 permission. Copyright 2016, WILEY-VCH. Figure 2c is reproduced with reference to 31 (a) permission. Copyright 2015, WILEY-VCH. Figure 2d is reproduced with reference to 31 (b) permission. Copyright 2017, American Chemical Society

potential generated by asymmetric metal contacts that effectively separate photogenerated electrons-hole pairs. So far, there have been many reports on BP-metal electrode Schottky contacts. However, it is rarely used in the field of photodetectors. Nevertheless, the excellent optical and electronic properties exhibited by the metal electrode-BP schottky heterojunction in the field of field effect transistors (Fig. 2b, c) [31] indicate that it has great potential applications in the field of photodetectors.

3 Black Phosphorous Heterojunction-Basde Photodetectors

Many previous reports have shown that BP photodetectors have excellent photoresponse performance and can achieve broadband photoresponse spectral range from ultraviolet to near infrared. However, the problem of plaguing all semiconductor materials, the recombination of photogenerated electron hole plays also limits the performance of BP photodetectors. At the same time, BP is easily oxidized in the air environment, which also limits the practical application of BP photodetectors. Subsequently, a method of constructing a hybrid heterojunction based on BP was proposed to enhance the stability and photoresponse performance of BP photodetectors. To date, many photodetectors based on BP heterojunctions have been reported and exhibit excellent photoresponse performance. This section has made a comprehensive summary of the research reports on BP heterojunction photodetectors in recent years. We classify them into three categories according to the structural characteristics of the heterojunction: 2D/2D heterojunction (vertical heterojunction, lateral heterojunction), mixed dimensional heterojunction (2D/3D heterojunction, 2D/1D heterojunction, 2D/0D heterojunction).

3.1 2D/2D Heterojunction Based on Black Phosphorous

BP is a typical 2D material with a unique structural advantage to form a 2D/2D heterojunction. The 2D/2D heterojunction is mainly divided into a vertical heterojunction and a lateral heterojunction. Among them, the 2D/2D vertical heterojunction is stacked by two 2D materials and combined by van der Waals force. The preparation process is simple and the performance is excellent, which makes the 2D/2D heterojunction widely studied and applied. As early as 2014, Deng et al. [32] demonstrated for the first time a BP/molybdenum disulfide (MoS_2) van der waals p-n heterojunction diode using P-type BP and n-type monolayer MoS_2. This ultra-thin p-n diode photodetector has a maximum photoresponsivity of 418 mA/W at a wavelength of 633 nm, which is nearly 100 times higher than the reported value of several layers of black phosphor phototransistors. Subsequently, Ye et al. [33] constructed a vertical heterojunction diode photodetector based on several layers of BP and a few-layers of MoS_2 (Fig. 3a). The photodetector has a detection range from visible to near-infrared light with a fast light response speed of 15 μs. Its photoresponsivity is 153.4 mA W^{-1} at $\lambda = 1.55$ μm, several orders of magnitude better than the reported value of a pure black phosphorus phototransistor. As the MoS_2/BP 2D/2D heterojunction exhibits excellent photoresponsive performance, more 2D materials are also used to construct

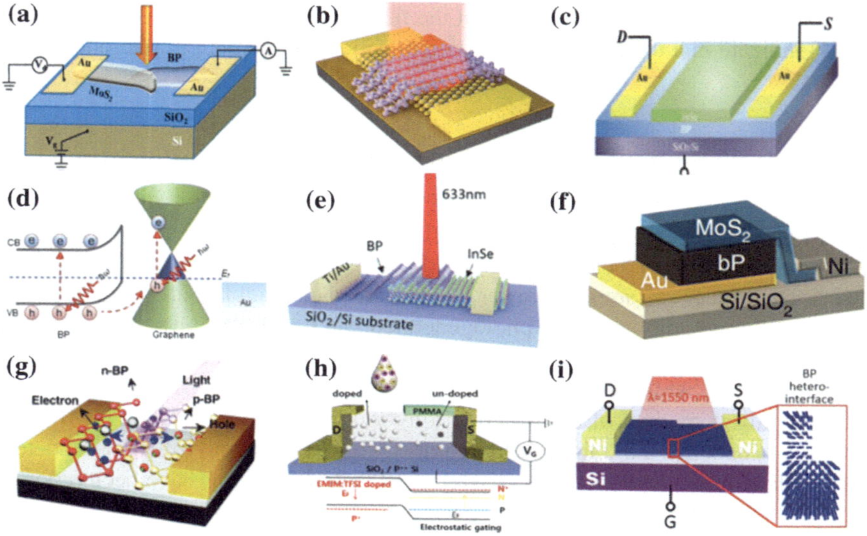

Fig. 3 **a** Schematic of the BP/MoS_2 heterojunction photodetector. **b** Schematic illustration of the BP-on-WSe_2 heterojunction photodetector. **c** Schematic illustration of the structure of BP/InSe heterojunction photoconductive detector. **d** Schematic diagram of graphene-BP heterostructure and photogenerated hot carrier transport under irradiation at zero gate voltage. **e** Schematic of the BP/InSe Vertical p–n heterojunction photodetector. **f** Schematic diagram of bP/MoS_2 homojunction photodiode. **g** Schematic diagram of Al-doped lateral p-n homojunction diod. **h** Schematic diagram of the lateral BP p-n junction device and the corresponding energy band diagram. **i** Schematic diagram of a thickness modulated BP heterojunction transistor. Figure 3a is reproduced with reference to 33 permission. Copyright 2016, American Chemical Society. Figure 3b is reproduced with reference to 34 permission. Copyright 2017, Elsevier Ltd. Figure 3c is reproduced with reference to 35 permission. Copyright 2019, WILEY-VCH. Figure 3d is reproduced with reference to 36 permission. Copyright 2017, American Chemical Society. Figure 3e is reproduced with reference to 37 permission. Copyright 2018, WILEY-VCH. Figure 3f is reproduced with reference to 38 permission. Copyright 2018, Springer Nature Publishing AG. Figure 3g is reproduced with reference to 39 permission. Copyright 2017, WILEY-VCH. Figure 3h is reproduced with reference to 40 permission. Copyright 2017, Springer Nature Publishing AG. Figure 3i is reproduced with reference to 41 permission. Copyright 2018, Royal Society of Chemistry

2D/2D vertical heterojunctions with BP. Such as, Ye et al. [34] constructed a broadband photodetector based on BP/WSe_2 vertical heterojunction and applied to infrared light detection (Fig. 3b). The photodetector has a light response of ~10^3 A/W and a light detection rate of ~10^{14} under visible light irradiation, and has a light response of ~5×10^{-1} A/W under infrared irradiation and ~10^{10} Light detection rate. This result is better than the BP photodetector and the WSe_2 photodetector under the same conditions. This is due to the built-in electric field of the WSe_2/BP heterojunction that promotes the efficient separation of photogenerated electron-hole pairs. In addition, in order to enhance the environmental stability of 2D BP, some special stacking 2D/2D vertical heterojunctions are designed. Gao et al. [35] fabricated a novel photoconductive detector based on several layers of BP/InSe vertical heterojunction on a SiO_2/Si substrate with a wide spectral detector range of 405–1550 nm (Fig. 3c). Under $\lambda = 655$ nm, P = 50 mW light irradiation,th e device exhibits a relatively fast response rate of 22 ms, a light responsiveness of ≈53.80 A W^{-1} and an external quantum efficiency of 1020%. Furthermore, vertical stacking of InSe nanosheets on top of BP can effectively inhibit BP degradation and enhance device environmental stability. Liu et al. [36] constructed a high performance infrared photodetector based on graphene-PB vertical heterojunction (Fig. 3d). At a near-infrared wavelength of 1550 nm, the graphene-BP vertical heterostructure phototransistor hasa high photoresponse of 3.3×10^3 AW^{-1}, a photoconductive gain of 1.13×109, and a fast response time of about 4 ms. This excellent photoresponse performance is due to the photogenerated electron-hole pairs in the BP being injected into the graphene, which significantly reduces the Schottky barrier between the BP and the metal electrode. This excellent photoresponse performance is due to the photogenerated electrons in the BP were inject into the graphene, which significantly reduces the schottky barrier between the BP and the metal electrode. At the same time, the coating of graphene enhances the environmental stability of BP, making the photodetector have a long-term stability.

For conventional photodetectors, it relies on an external electric field to separate photogenerated electron hole pairs. Therefore, it is necessary to design a self-powered photodetector that does not depend on an external electric field. It has been confirmed in the above reports that the p-n heterojunction can form a built-in electric field at the interface, thereby effectively promoting the separation of photogenerated electron hole pairs. Based on this fact, the p-n heterojunction is considered to be a potential structural model to construct self-powered photodetectors. Unfortunately, the previous report did not pay attention to this phenomenon. Zhao et al. [37] fabricated a vertical p-n diode photodetector by vertically stacking p-type few-layer BP on n-type few-layer InSe, which shows the optical response of broadband and gate modulation (Fig. 3e). In addition, the band alignment of BP/InSe p-n heterojunction can effectively promote the effective separation of photogenerated electron-hole pairs, which results in the device operating at zero bias voltage and achieving a fast photoresponse speed of approximately 10 ms, low dark current and up to 3% external quantum efficiency. This photoresponse characteristic under zero bias confirms that the BP/InSe p-n heterojunction can achieve self-powered light detection behavior. Furthermore, Bullock et al. [38] design a BP/MoS_2 vertical heterojunction photodiode and show excellent mid-wave infrared light detection performance (Fig. 3f). Compared to traditional BP diode photodetectors, the performance of the device has been significantly improved. At room temperature, the external quantum efficiency reaches 35%, and the detection rate is 1.1×10^{10} cm $Hz^{1/2}$ W^{-1} in the MWIR region. More interestingly, the device is capable of self-driven light detection without relying on external optics, with a fast photoresponse time of 3.7 μs at 0 bias.

2D lateral heterostructure as a typical 2D/2D heterojunction has unique structural features and provides theoretical and experimental guidance for the construction of heterojunction optoelectronic devices. However, the preparation process of this 2D/2D lateral heterojunction is very complicated. Under normal circumstances, lateral heterojunctions can only be prepared by a bottom-up approach, such as epitaxial growth. In addition, BP has poor environmental stability, and these factors have led to the recent study of BP-based 2D/2D lateral heterojunctions. Nevertheless, some BP lateral homojunction studies have been reported and applied to photodetection. Liu et al. [39] successfully prepared a few-layers BP p-n homojunctions by simple Al doping technique and constructed a few-layers BP p-n homojunction diodes (Fig. 3g). In the absence of external bias, the BP homojunction diode can still generate photovoltage and photocurrent to achieve self-powered photodetection behavior. The maximum open circuit photoresponsivity of the device is $\approx 15.7 \times 10^3$ V W^{-1}, the short-circuit photoresponsivity ≈ 6.2 mA W^{-1} is irradiated at a near-infrared 1550 nm wavelength and exhibits an optical on/off ratio higher than 10^3. Jia et al. [40] used a chemical doping method to form a uniform lateral p-n homojunction diode in a single piece of BP (Fig. 3h). The BP lateral p-n junction as a photodetector has a response time of less than 0.03 s, an external quantum efficiency of 2400%, and a light responsiveness of 120 A/W. This result is much larger than the recently reported layers of BP phototransistors and BP p-n junction photodetectors. Subsequently, Wang et al. [41] used an optical microscope to screen out BP sheets with thickness differences to construct an all-BP lateral heterojunction transistor (Fig. 3i). The band gap size of BP is controlled by its thickness, and BP can form a heterojunction between different thickness regions. Interestingly, the device has a high photoresponse of 383 A/W at l = 1550 nm and 0 V Vg. This shows that the device can achieve high performance self-powered light detection behavior without external voltage.

3.1.1 Mix-Dimensional Heterojunction Based on Black Phosphorous

As a 2D material, BP has many other forms of mixed-dimensional heterojunctions, including 2D/3D heterojunctions, 2D/1D heterojunctions, and 2D/0D heterojunctions. However, since the BP-based 2D/2D heterojunction exhibits extremely excellent photodetection performance, most of the studies on BP photodetection favor the 2D/2D heterojunction, and relatively few reports focusing on BP-based mix-dimensional heterojunctions. Here, we briefly summarize and review BP-based mix-dimensional heterojunction photodetectors.

The most common 2D/3D p-n heterojunction is the deposition of 2DBP nanosheets on a Si or GaAs bulk substrate to construct a 2D-3D p-n heterojunction photodetector. Gehring et al. [42] constructed a BP/GaAs p-n heterojunction diode by coating a small layer of p-type BP on a highly n-doped GaAs substrate, and evaluated its photoresponse characteristics (Fig. 4a). When irradiated with $\lambda = 514$ nm, the BP/GaAs p-n heterojunction has a photoreactivity of 37 mA/W and an external quantum efficiency of 9.7% at zero bias. The device exhibits excellent photoresponsive performance and enables self-powered photodetector behavior independent of external voltage. In addition, some studies have pointed out that atom and ion doped BP can enhance its photoresponse performance and environmental stability. Xu et al. [43] reported a Se-doped BP field-effect transistor in which a centimeter-Se-doped BP crystal was successfully prepared by vapor phase growth, and a small layer of

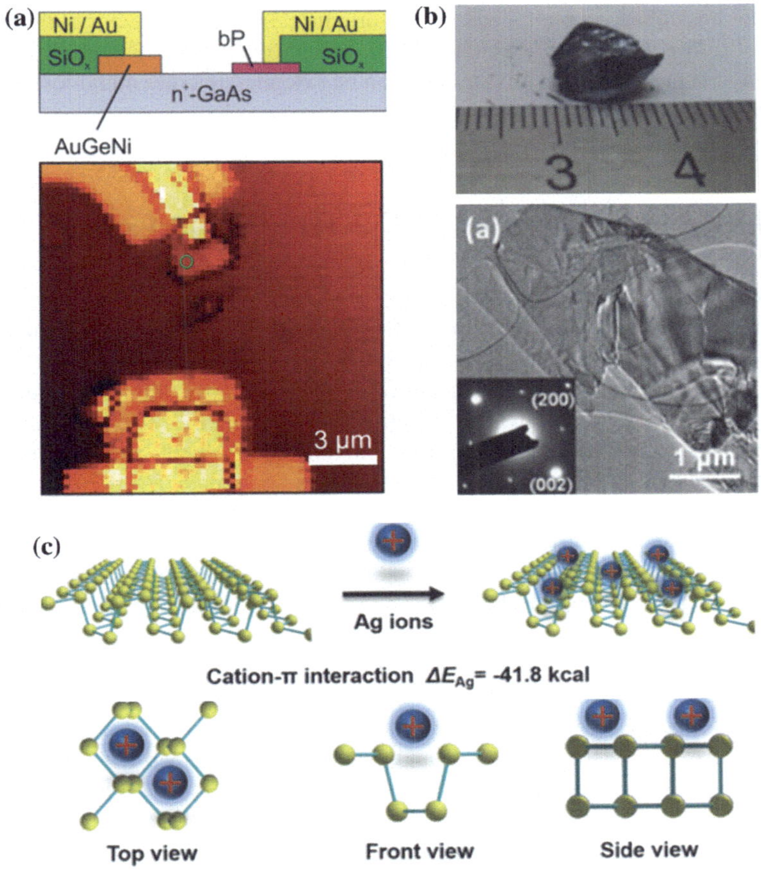

Fig. 4 **a** 2D/3D black phosphor/GaAs p-n heterojunction diode schematic and optical reflection image. **b** Optical image of a centimeter-sized Se-doped BP crystal and TEM image of a Se-doped BP flake. **c** Schematic diagram of adsorption of Ag^+ on BP nanosheets. Figure 4a is reproduced with reference to 42 permission. Copyright 2015, AIP Publishing LLC. Figure 4b is reproduced with reference to 43 permission. Copyright 2016, Wiley-VCH. Figure 4c is reproduced with reference to 44 permission. Copyright 2017, Wiley-VCH

Se-doped BP nanosheets was obtained by mechanical exfoliation (Fig. 4b). This few-layers of Se-doped BP field effect transistor has a high on/off current ratio of 10^5, an optical responsivity of 15.33 AW^{-1} and an external quantum efficiency of 2993%. Compared to a single layerless BP photodetector, its optical responsiveness is increased by 20 times. Gao et al. [44] reported a metal ion-modified BP nanosheet field effect transistor that enhances the environmental stability of BP by surface Ag^+ adsorption (Fig. 4c). The Ag-modified BP transistor has enhanced stability because free Ag^+ is adsorbed on the BP surface to passivate the lone-pair electrons of the P atom, thereby inhibiting the degradation of BP in the air. At the same time, the transistor has enhanced photodetector performance as a photodetector with an optical on/off ratio of 2.6×10^6, which is 44 times that of a single BP photodetector. In addition, it is predicted that this strategy can be extended to other metal ions such as Fe^{3+}, Mg^{2+} and Hg^{2+}.

4 Black Phosphorous-Based PEC–Type Photodetectors

PEC-type photodetectors is a new type of photodetector based on PEC system. Compared with traditional photodetectors, PEC photodetectors have the advantages of simple fabrication process and low cost. At the same time, PEC photodetectors have greater photocurrent. For most conventional photodetectors, the output current is typically on the order of nA, while the output current of a PEC photodetector can be on the order of μA [6a]. In addition, PEC photodetectors enable self-powered photodetector behavior without relying on external power supplies. The performance test of the PEC-type photodetector is through a three-electrode electrochemical workstation, which includes a working electrode (sample coated on the conductive surface of the Fluorine-doped Tin Oxide/Indium Tin Oxide (FTO/ITO) glass as a photoanode), a counter electrode (platinum sheet), and a reference electrode (Ag/AgCl electrode). The working mechanism of the PEC type photodetector are shown in Fig. 5a, b [45]. Under illumination conditions, the photoanode material produces photogenerated electron-hole pairs under illumination, wherein electrons are transferred to the counter electrode (platinum plate) via the ITO conductive substrate and combined with H^+ ions in the electrolyte, and the holes are combined with anions in the electrolyte. In this process, the electrolyte exchanges electrons with photogenerated electron-hole pairs and acts as an ion channel to complete the entire current loop. This section is mainly to summarize and review latest developments of BP in the field of PEC photodetectors.

In recent years, the development of PEC photodetectors has been rapid, and many semiconductor materials have been used as photoanode materials for PEC photodetectors, such as InSe [46], TiO_2 [45b, 47], ZnO [48], Bi_2S_3 [49], 2D Bi nanosheets [50], and exhibit excellent optical response properties. However, there are few studies on BP-based PEC-type photodetectors. Recently, Ren et al. [51] constructed a self-powered PEC photodetector in KOH solution based on liquid exfoliated few-layer BP nanosheets (Fig. 5c). At the voltage is zero, the PEC-type BP photodetector has a photocurrent density of 265 nA cm^{-2} under light irradiation, while the dark current density of 1 nA cm^{-2} in 0.1 M KOH. This indicates that the device has self-powered light detection characteristics with a photoresponsivity of 2.2 μAW^{-1} and an optical on/off ratio of approximately 265. Besides this, the PEC-type BP photodetector has good stability and the device still maintains a good on/off behavior after 1 month. The reason for this is because OH^- ions in the alkaline electrolyte can be adsorbed on the BP surface and generate a negative charge. The presence of these negative charges can effectively inhibit the oxidation process of BP and enhance its environmental stability [17, 52]. This PEC-type BP photodetector provides new ideas for

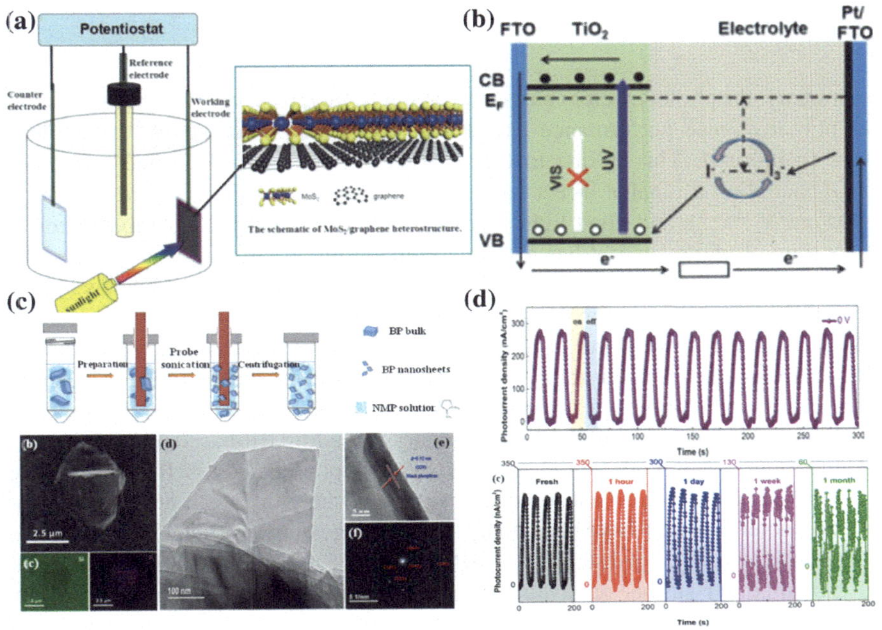

Fig. 5 **a** Schematic diagram of the PEC-type photodetector. **b** Schematic diagram of electron transfer process of PEC-type photodetector. **c** Schematic diagram of liquid phase stripping of BP nanosheets. **d** SEM, EDS, TEM, HRTEM characterization images of exfoliated BP nanosheets. Figure 5a is reproduced with reference to 45 (a) permission. Copyright 2015, IOP Publishing Ltd. Figure 5b is reproduced with reference to 45 (b) permission. Copyright 2012, Elsevier Ltd. Figure 5c, d are reproduced with reference to 51 permission. Copyright 2017, Wiley-VCH

developing high responsivity, low cost, and easy to manufacture self-powered photodetectors. Although BP-based PEC-type photodetectors have been less studied, the excellent photoresponse performance exhibited by PEC-type BP photodetectors indicates that BP has a good prospect in the field of PEC-type photodetectors. For example, the application of BP hybrid heterojunction to PEC photodetectors is an effective method for developing high-response, high-stability self-powered photodetectors. At present, a new type of solid electrolyte has been reported and widely used in photoelectrochemical systems. This provides theoretical guidance for the further development of the PEC-type BP photodetector, which is to build a high performance, high stability flexible, wearable, self-powered PEC-type BP photodetector.

5 Summary

In summary, the photodetectors based on BP and BP heterojunctions exhibited excellent photoresponse performance and a wide range of photoresponse from ultraviolet to infrared. This is due to the excellent optical and electronic properties of BP, including a tunable direct bandgap (0.3–2 eV), excellent electron mobility (1000 $cm^2V^{-1}s^{-1}$), and good light absorption. In particular, PEC photodetector as a new type of photodetector has a simple preparation process and can realize self-powered photodetection behavior for all photosensitive materials, which has important guiding significance for the design and development of new devices. In this chapter, we briefly review the structural features, electrical and optical properties of BP. In addition, the research status of BP-based photodetectors is introduced in detail, and its further development is forecasted. We divided the BP photodetectors into three categories and made a comprehensive summary, including BP-based photodetectors, BP heterojunction photodetectors, and PEC-type BP photodetectors. Besides this, we also focused on the review and summary of BP nanosheets in the air environment for easy degradation. There have been many reports suggesting several effective ways to improve the environmental stability of BP nanosheets, such as encapsulation and isolation of air, or BP surface passivation, elemental doping and metal ion modification. We hope that this chapter will provide readers with a comprehensive understanding of the research status of BP photodetectors and provide positive guidance for further design and development of high performance photodetectors.

Acknowledgements This work was supported by the Provincial Natural Science Foundation of Hunan (No. 2016JJ2132), Science and Technology Program of Xiangtan (No. CXY-ZD20172002), and the Program for Changjiang Scholars and Innovative Research Team in University (IRT_17R91)

References

1. (a) Xia, F., Mueller, T., Lin, Y.-m., Valdes-Garcia, A., Avouris, P.: Nat. Nanotechnol. **4**, 839 (2009); (b) Gan, X., Shiue, R.-J., Gao, Y., Meric, I., Heinz, T.F., Shepard, K., Hone, J., Assefa, S., Englund, D.: Nat. Photon. **7**, 883 (2013); (c) Lee, Y., Kwon, J., Hwang, E., Ra, C.H., Yoo, W.J., Ahn, J.H., Park, J.H., Cho, J.H.: Adv. Mater. **27**, 41 (2015); (d) Wang, B., Zhong, S.P., Zhang, Z.B., Zheng, Z.Q., Zhang, Y.P., Zhang, H.: Appl. Mater. Today **15**, 115 (2019); (e) Fan, P., Huang, K.C., Cao, L., Brongersma, M.L.: Nano Lett. **13**, 392 (2013)
2. (a) Bao, C., Yang, J., Bai, S., Xu, W., Yan, Z., Xu, Q., Liu, J., Zhang, W., Gao, F.: Adv. Mater. **30**, 1803422 (2018); (b) Pospischil, A., Humer, M., Furchi, M.M., Bachmann, D., Guider, R., Fromherz, T., Mueller, T., Nat. Photon. **7**, 892 (2013); (c) Kharraz, O., Forsyth, D.: Optik-Int. J. Light Electron Opt. **124**, 1493 (2013); (d) Arredondo, B., Romero, B., Pena, J., Fernández-Pacheco, A., Alonso, E., Vergaz, R., de Dios, C.: Sensors **13**, 12266 (2013)
3. Yang, Q., Liu, Y., Li, Z., Yang, Z., Wang, X., Wang, Z.L.: Angew. Chem. Int. Ed. **51**, 6443 (2012)
4. (a) LaPierre, R., Robson, M., Azizur-Rahman, K., Kuyanov, P.: J. Phys. D: Appl. Phys. **50**, 123001 (2017); (b) Luxmoore, I.J., Liu, P.Q., Li, P., Faist, J., Nash, G.R.: ACS Photon. **3**, 936

(2016); (c) Park, S., Fukuda, K., Wang, M., Lee, C., Yokota, T., Jin, H., Jinno, H., Kimura, H., Zalar, P., Matsuhisa, N.: Adv. Mater. 30, 1802359 (2018)
5. Son, K.T., Lee, C.C.: IEEE Trans. Instrum. Meas. **59**, 3005 (2010)
6. (a) Zhou, J., Chen, L., Wang, Y., He, Y., Pan, X., Xie, E.: Nanoscale **8**, 50 (2016); (b) Ahmadi, M., Wu, T., Hu, B.: Adv. Mater. **29**, 1605242 (2017); (c) Barve, A.V., Lee, S.J., Noh, S.K., Krishna, S.: Laser Photon. Rev. **4**, 738 (2010)
7. (a) Xie, C., Mak, C., Tao, X., Yan, F.: Adv. Funct. Mater. **27**, 1603886 (2017); (b) Lopez-Sanchez, O., Lembke, D., Kayci, M., Radenovic, A., Kis, A.: Nat. Nanotechnol. **8**, 497 (2013); (c) Koppens, F., Mueller, T., Avouris, P., Ferrari, A., Vitiello, M., Polini, M.: Nat. Nanotechnol. **9**, 780 (2014); (d) Mueller, T., Xia, F., Avouris, P.: Nat. Photon. **4**, 297 (2010)
8. Bridgman, P.: J. Am. Chem. Soc. **36**, 1344 (1914)
9. (a) Koenig, S.P., Doganov, R.A., Schmidt, H., Castro Neto, A., Oezyilmaz, B.: Appl. Phys. Lett. **104**, 103106 (2014); (b) Tran, V., Soklaski, R., Liang, Y., Yang, L.: Phys. Rev. B **89**, 235319 (2014); (c) Castellanos-Gomez, A.: J. Phys. Chem. Lett. **6**, 4280 (2015); (d) Deng, B., Tran, V., Xie, Y., Jiang, H., Li, C., Guo, Q., Wang, X., Tian, H., Koester, S.J., Wang, H.: Nat. Commun. **8**, 14474 (2017)
10. (a) Yuan, H., Liu, X., Afshinmanesh, F., Li, W., Xu, G., Sun, J., Lian, B., Curto, A.G., Ye, G., Hikita, Y.: Nat. Nanotechnol. **10**, 707 (2015); (b) Xia, F., Wang, H., Jia, Y.: Nat. Commun. **5**, 4458 (2014)
11. Li, L., Yu, Y., Ye, G.J., Ge, Q., Ou, X., Wu, H., Feng, D., Chen, X.H., Zhang, Y.: Nat. Nanotechnol. **9**, 372 (2014)
12. (a) Island, J.O., Steele, G.A., van der Zant, H.S., Castellanos-Gomez, A.: 2D Mater. **2**, 011002 (2015); (b) Wan, B., Yang, B., Wang, Y., Zhang, J., Zeng, Z., Liu, Z., Wang, W.: Nanotechnology **26**, 435702 (2015); (c) Xing, C., Jing, G., Liang, X., Qiu, M., Li, Z., Cao, R., Li, X., Fan, D., Zhang, H.: Nanoscale **9**, 8096 (2017)
13. Zhao, Y., Wang, H., Huang, H., Xiao, Q., Xu, Y., Guo, Z., Xie, H., Shao, J., Sun, Z., Han, W.: Angew. Chem. Int. Ed. **55**, 5003 (2016)
14. Illarionov, Y.Y., Waltl, M., Rzepa, G., Kim, J.-S., Kim, S., Dodabalapur, A., Akinwande, D., Grasser, T.: ACS Nano **10**, 9543 (2016)
15. Shao, J., Xie, H., Huang, H., Li, Z., Sun, Z., Xu, Y., Xiao, Q., Yu, X.-F., Zhao, Y., Zhang, H.: Nat. Commun. **7**, 12967 (2016)
16. Kang, J., Wood, J.D., Wells, S.A., Lee, J.-H., Liu, X., Chen, K.-S., Hersam, M.C.: ACS Nano **9**, 3596 (2015)
17. Xu, Y., Wang, Z., Guo, Z., Huang, H., Xiao, Q., Zhang, H., Yu, X.F.: Adv. Opt. Mater. **4**, 1223 (2016)
18. (a) Zou, J., Zhang, Q., Huang, K., Marzari, N.: J. Phys. Chem. C **114**, 10725 (2010); (b) Xie, Y., Wei, L., Li, Q., Chen, Y., Yan, S., Jiao, J., Liu, G., Mei, L.: Nanotechnology **25**, 075202 (2014); (c) Ellis, J.K., Lucero, M.J., Scuseria, G.E.: Appl. Phys. Lett. **99**, 261908 (2011)
19. (a) Zhang, M., Wu, Q., Zhang, F., Chen, L., Jin, X., Hu, Y., Zheng, Z., Zhang, H.: Adv. Opt. Mater. **7**, 1800224 (2019); (b) Low, T., Rodin, A., Carvalho, A., Jiang, Y., Wang, H., Xia, F., Neto, A.C.: Phys. Rev. B **90**, 075434 (2014)
20. (a) Qiao, J., Kong, X., Hu, Z.-X., Yang, F., Ji, W.: Nat. Commun. **5**, 4475 (2014); (b) Fei, R., Yang, L.: Nano Lett. **14**, 2884 (2014); (c) Low, T., Engel, M., Steiner, M., Avouris, P.: Physical Review B **90**, 081408 (2014)
21. (a) Long, M., Gao, A., Wang, P., Xia, H., Ott, C., Pan, C., Fu, Y., Liu, E., Chen, X., Lu, W.: Sci. Adv. **3**, e1700589 (2017); (b) Chen, X., Lu, X., Deng, B., Sinai, O., Shao, Y., Li, C., Yuan, S., Tran, V., Watanabe, K., Taniguchi, T.: Nat. Commun. **8**, 1672 (2017)
22. Engel, M., Steiner, M., Avouris, P.: Nano Lett. **14**, 6414 (2014)
23. Guo, Q., Pospischil, A., Bhuiyan, M., Jiang, H., Tian, H., Farmer, D., Deng, B., Li, C., Han, S.-J., Wang, H.: Nano Lett. **16**, 4648 (2016)
24. Wu, J., Koon, G.K.W., Xiang, D., Han, C., Toh, C.T., Kulkarni, E.S., Verzhbitskiy, I., Carvalho, A., Rodin, A.S., Koenig, S.P.: ACS Nano **9**, 8070 (2015)
25. Youngblood, N., Chen, C., Koester, S.J., Li, M.: Nat. Photon. **9**, 247 (2015)
26. Huang, M., Wang, M., Chen, C., Ma, Z., Li, X., Han, J., Wu, Y.: Adv. Mater. **28**, 3481 (2016)

27. (a) Favron, A., Gaufrès, E., Fossard, F., Phaneuf-L'Heureux, A.-L., Tang, N.Y., Lévesque, P.L., Loiseau, A., Leonelli, R., Francoeur, S., Martel, R.: Nat. Mater. **14**, 826 (2015); (b) Doganov, R.A., O'Farrell, E.C., Koenig, S.P., Yeo, Y., Ziletti, A., Carvalho, A., Campbell, D.K., Coker, D.F., Watanabe, K., Taniguchi, T.: Nat. Commun. **6**, 6647 (2015); (c) Kim, J.-S., Liu, Y., Zhu, W., Kim, S., Wu, D., Tao, L., Dodabalapur, A., Lai, K., Akinwande, D.: Sci. Rep. **5**, 8989 (2015)
28. Wood, J.D., Wells, S.A., Jariwala, D., Chen, K.-S., Cho, E., Sangwan, V.K., Liu, X., Lauhon, L.J., Marks, T.J., Hersam, M.C.: Nano Lett. **14**, 6964 (2014)
29. Na, J., Park, K., Kim, J.T., Choi, W.K., Song, Y.-W.: Nanotechnology **28**, 085201 (2017)
30. Miao, J., Zhang, S., Cai, L., Wang, C.: Adv. Electron. Mater. **2**, 1500346 (2016)
31. (a) Ling, Z.-P., Sakar, S., Mathew, S., Zhu, J.-T., Gopinadhan, K., Venkatesan, T., Ang, K.-W.: Sci. Rep. **5**, 18000 (2015); (b) Quhe, R., Peng, X., Pan, Y., Ye, M., Wang, Y., Zhang, H., Feng, S., Zhang, Q., Shi, J., Yang, J.: ACS Appl. Mater. Interfaces **9**, 3959 (2017)
32. Deng, Y., Luo, Z., Conrad, N.J., Liu, H., Gong, Y., Najmaei, S., Ajayan, P.M., Lou, J., Xu, X., Ye, P.D.: ACS Nano **8**, 8292 (2014)
33. Ye, L., Li, H., Chen, Z., Xu, J.: Acs Photon. **3**, 692 (2016)
34. Ye, L., Wang, P., Luo, W., Gong, F., Liao, L., Liu, T., Tong, L., Zang, J., Xu, J., Hu, W.: Nano Energy **37**, 53 (2017)
35. Cao, R., Wang, H.D., Guo, Z.N., Sang, D.K., Zhang, L.Y., Xiao, Q.L., Zhang, Y.P., Fan, D.Y., Li, J.Q., Zhang, H.: Adv. Opt. Mater. 1900020 (2019)
36. Liu, Y., Shivananju, B.N., Wang, Y., Zhang, Y., Yu, W., Xiao, S., Sun, T., Ma, W., Mu, H., Lin, S.: ACS Appl. Mater. Interfaces **9**, 36137 (2017)
37. Zhao, S., Wu, J., Jin, K., Ding, H., Li, T., Wu, C., Pan, N., Wang, X.: Adv. Funct. Mater. **28**, 1802011 (2018)
38. Bullock, J., Amani, M., Cho, J., Chen, Y.-Z., Ahn, G.H., Adinolfi, V., Shrestha, V.R., Gao, Y., Crozier, K.B., Chueh, Y.-L.: Nat. Photon. **12**, 601 (2018)
39. Liu, Y., Cai, Y., Zhang, G., Zhang, Y.W., Ang, K.W.: Adv. Funct. Mater. **27**, 1604638 (2017)
40. Jia, J., Xu, J., Park, J.-H., Lee, B.H., Hwang, E., Lee, S.: Chem. Mater. **29**, 3143 (2017)
41. Wang, L., Huang, L., Tan, W.C., Feng, X., Chen, L., Ang, K.-W.: Nanoscale **10**, 14359 (2018)
42. Gehring, P., Urcuyo, R., Duong, D.L., Burghard, M., Kern, K.: Appl. Phys. Lett. **106**, 233110 (2015)
43. Xu, Y., Yuan, J., Fei, L., Wang, X., Bao, Q., Wang, Y., Zhang, K., Zhang, Y.: Small **12**, 5000 (2016)
44. Guo, Z., Chen, S., Wang, Z., Yang, Z., Liu, F., Xu, Y., Wang, J., Yi, Y., Zhang, H., Liao, L.: Adv. Mater. **29**, 1703811 (2017)
45. (a) Huang, Z., Han, W., Tang, H., Ren, L., Chander, D.S., Qi, X., Zhang, H.: 2D Mater. **2**, 035011 (2015); (b) Li, X., Gao, C., Duan, H., Lu, B., Pan, X., Xie, E., Nano Energy **1**, 640 (2012)
46. Li, Z., Qiao, H., Guo, Z., Ren, X., Huang, Z., Qi, X., Dhanabalan, S.C., Ponraj, J.S., Zhang, D., Li, J.: Adv. Funct. Mater. **28**, 1705237 (2018)
47. Li, X., Gao, C., Duan, H., Lu, B., Wang, Y., Chen, L., Zhang, Z., Pan, X., Xie, E.: Small **2013**, 9 (2005)
48. Majumder, T., Hmar, J., Debnath, K., Gogurla, N., Roy, J., Ray, S., Mondal, S.: J. Appl. Phys. **116**, 034311 (2014)
49. Huang, W., Xing, C., Wang, Y., Li, Z., Wu, L., Ma, D., Dai, X., Xiang, Y., Li, J., Fan, D.: Nanoscale **10**, 2404 (2018)
50. Huang, H., Ren, X., Li, Z., Wang, H., Huang, Z., Qiao, H., Tang, P., Zhao, J., Liang, W., Ge, Y.: Nanotechnology **29**, 235201 (2018)
51. Ren, X., Li, Z., Huang, Z., Sang, D., Qiao, H., Qi, X., Li, J., Zhong, J., Zhang, H.: Adv. Funct. Mater. **27**, 1606834 (2017)
52. Guo, Z., Zhang, H., Lu, S., Wang, Z., Tang, S., Shao, J., Sun, Z., Xie, H., Wang, H., Yu, X.F.: Adv. Funct. Mater. **25**, 6996 (2015)

Correction to: Black Phosphorus

Inamuddin, Rajender Boddula and Abdullah M. Asiri

Correction to:
Inamuddin et al. (eds.), *Black Phosphorus*, Engineering Materials, https://doi.org/10.1007/978-3-030-29555-4

This book was inadvertently published with incorrect affiliation and updated in the book.

Inamuddin, Department of Chemistry, Faculty of Science, King Abdulaziz University, Jeddah, Saudi Arabia.

Rajender Boddula, CAS Key Laboratory of Nanosystem and Hierarchical Fabrication, National Center for Nanoscience and Technology, Beijing, China.

Abdullah M. Asiri, Department of Chemistry, Faculty of Science, King Abdulaziz University, Jeddah, Saudi Arabia.

The updated version of the book can be found at
https://doi.org/10.1007/978-3-030-29555-4

Inamuddin et al. (eds.), *Black Phosphorus*, Engineering Materials,
https://doi.org/10.1007/978-3-030-29555-4_10

The manufacturer's authorised representative in the EU is Springer Nature Customer Service Centre GmbH, Europaplatz 3, 69115 Heidelberg, Germany. If you have any concerns regarding our products, please contact ProductSafety@springernature.com

Printed and bound by CPI Group (UK) Ltd, Croydon, CR0 4YY
22/09/2026
02221060-0001